MW00760512

Springer Theses

Recognizing Outstanding Ph.D. Research

Aims and Scope

The series "Springer Theses" brings together a selection of the very best Ph.D. theses from around the world and across the physical sciences. Nominated and endorsed by two recognized specialists, each published volume has been selected for its scientific excellence and the high impact of its contents for the pertinent field of research. For greater accessibility to non-specialists, the published versions include an extended introduction, as well as a foreword by the student's supervisor explaining the special relevance of the work for the field. As a whole, the series will provide a valuable resource both for newcomers to the research fields described, and for other scientists seeking detailed background information on special questions. Finally, it provides an accredited documentation of the valuable contributions made by today's younger generation of scientists.

Theses are accepted into the series by invited nomination only and must fulfill all of the following criteria

- They must be written in good English.
- The topic should fall within the confines of Chemistry, Physics, Earth Sciences, Engineering and related interdisciplinary fields such as Materials, Nanoscience, Chemical Engineering, Complex Systems and Biophysics.
- The work reported in the thesis must represent a significant scientific advance.
- If the thesis includes previously published material, permission to reproduce this must be gained from the respective copyright holder.
- They must have been examined and passed during the 12 months prior to nomination.
- Each thesis should include a foreword by the supervisor outlining the significance of its content.
- The theses should have a clearly defined structure including an introduction accessible to scientists not expert in that particular field.

More information about this series at http://www.springer.com/series/8790

Diana Martín Becerra

Active Plasmonic Devices

Based on Magnetoplasmonic Nanostructures

Doctoral Thesis accepted by
Complutense University of Madrid, Spain

Springer

Author
Dr. Diana Martín Becerra
Instituto de Microelectrónica de Madrid
Madrid
Spain

Supervisors
Prof. Maria Ujué González Sagardoy
Instituto de Microelectrónica de Madrid
Madrid
Spain

Prof. Antonio García-Martín
Instituto de Microelectrónica de Madrid
Madrid
Spain

ISSN 2190-5053 ISSN 2190-5061 (electronic)
Springer Theses
ISBN 978-3-319-48410-5 ISBN 978-3-319-48411-2 (eBook)
DOI 10.1007/978-3-319-48411-2

Library of Congress Control Number: 2016955326

Printed on acid-free paper

This Springer imprint is published by Springer Nature
The registered company is Springer International Publishing AG
The registered company address is: Gewerbestrasse 11, 6330 Cham, Switzerland

A los que están,
y a los que
ya se han ido

Supervisors' Foreword

Since the beginning of the present century, it has been possible to overcome the diffraction limit and confine light in subwavelength volumes, enabling important progress in the development of nanophotonic devices and structures. One of the most frequently exploited routes has been the use of metallic nanostructures, taking the advantage of the excitation of surface plasmons (SP), collective electron oscillations occurring at the interface between a metal and a dielectric. This subfield of nanophotonics, denoted plasmonics, has subsequently evolved in many varied directions and fostered a large number of applications. For example, the above-mentioned ability of SPs to confine the electromagnetic light field in small volumes results in huge local field enhancements. This, in turn, increases the interaction of light with molecules, making plasmonics widely used in molecule detection through SERS (surface-enhanced Raman scattering). Since the properties of SPs are highly sensitive to the optical properties (refractive index) of the dielectric media surrounding the metal, sensing has become one of the main and better-established applications in the field. Additionally, SPs can be guided and manipulated at the surface of nanostructured metals; thus, they are also being employed to build nanophotonic chips for communications and signal processing.

Within the context of the development of plasmonic circuits, plasmonic waveguides were easily obtained, but the realization of fundamental components in nanophotonic chips, such as modulators, switches or active multiplexors, and couplers, requires finding ways to control SP properties via external agents. In recent years, active plasmonic configurations have been achieved based on the different controlling mechanisms, such as temperature, voltage, or light. The main drawback of most methods is the slow response of the material to the stimuli, preventing the accomplishment of high switching speeds. Light has been, so far, the only option able to overcome this problem. A competitive alternative candidate for getting an active response in a plasmonic system is the use of a magnetic field, as it allows a modification of the optical properties of a broad range of materials including metals. Moreover, as magnetism is intrinsically an ultrafast property, modulation speeds at femtosecond levels could be attainable.

The combination of magnetic and plasmonic functionalities in different nanostructures has latterly become an active topic of research, referred to as magnetoplasmonics. The interaction between the two effects is bidirectional: On the one hand, the field enhancement associated with the SP excitation can increase the magneto-optical response, and on the other, the magnetic field can modify the properties of the SPs. In order to obtain a significant interaction between the two effects, materials with good plasmonic and good magnetic properties have to be merged. By fabricating multilayers of noble and ferromagnetic metals, the modification of the wavevector of the plasmons propagating along the metal surface by means of an applied magnetic field has been demonstrated. Moreover, by engraving these multilayers with nanoslits and grooves, magnetoplasmonic interferometers were obtained that can act as optical modulators.

During her Ph.D., Diana Martín-Becerra undertook a systematic study of this kind of magnetoplasmonic interferometer. She has employed them to determine the magnitude of the magnetic modulation of the SP wavevector for different material systems, multilayer configurations, and wavelengths. This analysis has allowed her to understand the mechanisms and properties governing the response of the system and to discuss the feasibility of using the interferometers as active plasmonic devices in a competitive way. One interesting finding has been that the magnitude of the magnetic field induced modification of the SP wavevector has a quadratic dependence on the permittivity of the dielectric material at the interface with the metal. Beyond providing a strategy for designing interferometers with a higher response, this indicates that this magnitude can be employed as a sensing parameter. Therefore, Diana has also evaluated the exploitation of magnetoplasmonic interferometers for sensing and compared their performance with that of the mainly used surface plasmon resonance (SPR)-based technique. The combination of detailed background information, discussion, and new results presented in this thesis will make it useful and stimulating reading, especially for newcomers to plasmonics interferometry, active plasmonics, and magnetoplasmonics.

Tres Cantos, Madrid, Spain
September 2016

Prof. María Ujué González Sagardoy
Prof. Antonio García-Martín

Abstract

Surface plasmons (SPs) constitute a promising route toward the development of miniaturized optical devices. Several passive plasmonic systems have been successfully demonstrated in the last decade, but the achievement of nanophotonic devices with advanced functionalities requires the implementation of active configurations. Among the different control agents considered so far to manipulate the SPs, the magnetic field is a strong candidate since it is able to directly modify the dispersion relation of SPs (Wallis et al. in Phys. Rev. B 9:3424–3437, 1974), with a high switching speed. This modification lies on the non-diagonal elements of the dielectric tensor, ε_{ij}. For noble metals, the ones typically used in plasmonics, these elements are unfortunately very small at reasonable field values. On the other hand, ferromagnetic metals have large ε_{ij} values at small magnetic fields (proportional to their magnetization), but they are optically too absorbent. Thereby, a smart system to develop magnetic field-sensitive plasmonic devices could be multilayers of noble and ferromagnetic metals (Gonzalez-Diaz et al. in Phys. Rev. B 76:153402, 2007; Ferreiro-Vila et al. in Phys. Rev. B 83:205120, 2011). Magnetoplasmonic modulation of surface plasmon polariton (SPP) wavevector in these hybrid Au/Co/Au multilayer films has been recently demonstrated (Temnov et al. in Nat. Photonics 4:107–111, 2010). These magnetoplasmonic modulators are based on the plasmonic microinterferometers formed by a tilted slit–groove pair. When illuminating the interferometers with a polarized laser, the light collected at the other side of the slit consists of the interference between the light directly transmitted through the slit and a SPP excited in the groove and reconverted back to radiative light in the slit. When we apply an external oscillating magnetic field, both the real and imaginary parts of the SPP wavevector are modified, therefore changing the interference intensity synchronously with the applied magnetic field. The intensity modulation depth of the system is given by the product $\Delta k_x^r \times d$, where Δk_x^r is the real part of the SPP wavevector modification induced by the magnetic field and d id the groove–slit distance. On the other hand, the imaginary part of the SPP wavevector modification introduces a phase shift, which is quite small and has not been done before, and it is related to the propagation distance of the SPP. A spectral analysis of those two parameters (Δk_x^r and Δk_x^i) has been done to have a better characterization of the modulation. However, the modulation of the real part of the wavevector obtained in this basic configuration (Au/Co/Au multilayers in air) does not exceed a few percent (Temnov et al. in Nat. Photonics 4:107–111, 2010), which is not strong enough for practical applications. We present a straightforward approach to increase the modulation of the SPP wavevector by covering the metallic multilayer with a dielectric media possessing a higher permittivity d. With our magnetoplasmonic interferometers covered by a thin layer of PMMA ($\varepsilon_d = 2.22$), we have obtained a fourfold enhancement of the Δk_x value, in excellent agreement with theoretical predictions (Martin-Becerra et al. in Appl. Phys. Lett. 97:183114, 2010). However, the propagation distance of the plasmon, L_{sp}, decreases with the addition of dielectric overlayers, which will prevent the use of interferometers with large d limiting the intensity modulation depth. The relevant figure of merit in this case is the product $\Delta k_x \times L_{sp}$. Our theoretical results show that with an optimized thickness of a polymer cover layer, this product can be almost doubled. A detailed analysis of the behavior of magnetoplasmonic interferometers covered with dielectric overlayers, in terms of both modulation enhancement and propagation distance, will be presented. On the other hand, the strong dependence of Δk_x with ε_d suggests that this magnitude could have sensing capabilities. A theoretical analysis of the sensitivity of this parameter, compared together with the typically used plasmon sensing SPR-based technique (Maier in Plasmonics: Fundamentals and Applications, Springer, 2007; Raether in Surface Plasmons, Springer, 1986), will be presented. Our results show that, at 633 nm wavelength, the sensitivity of an interferometric configuration can be higher than that of the SPR one because of the slit–groove distance role. Finally, due to the local nature of SPs, we have analyzed the effect of applying a magnetic field on propagating SP in the near-field regime. Some alternative interferometric plasmonic configurations proposed in the literature for near-field studies (Wang et al. in Appl. Phys. Lett. 94:153902, 2007; Aigouy et al. in Phys. Rev. Lett. 98:153902, 2007) have been considered.

List of Publications

Publications directly related to this thesis are as follows:

- Martín-Becerra, D.; González-Díaz, J.B.; Temnov, V.V.; Cebol lada, A.; Armelles, G.; Thomay, T.; Leitenstorfer, A.L.; Bratschitsch, R.; García-Martín, A. and González, M.U.
  Enhancement of the magnetic modulation of surface plasmon polaritons in Au/Co/Au films. Appl. Phys. Lett., 2010, 97, 183114
- Martín-Becerra, D.; Temnov, V.V.; Thomay, T.; Leitenstorfer, A.; Bratschitsch, R.; Armelles, G.; García-Martín, A. and Gonzĺez, M. U.
  Spectral dependence of the magnetic modulation of surface plasmon polaritons in noble/ferromagnetic/noble metal films. Phys. Rev. B, American Physical Society, 2012, 86, 035118
- Martín-Becerra, D.; Armelles, G.; González, M.U. and García-Martín, A.
  Plasmonic and magnetoplasmonic interferometry for sensing. New J. Phys., 2013, 15, 085021
- Martín-Becerra, D.; García-Martín, A.; and González, M.U.
  Magnetoplasmonics: magneto-optical material dielectric or metal. (in preparation)

Other publications

- Martín-Becerra, D.; García-Martín, J.M.; Huttel, Y.; and Armelles, G.
  Optical and magneto-optical properties of Au: Co nanoparticles and Co: Au nanoparticles doped magnetoplasmonic systems, JAP, 2015, 117, 053101
- Kekesi, R. and Martín-Becerra, D. and Meneses-Rodríguez, D. and García-Pérez, F. and Cebollada, A. and Armelles, G.
  Enhanced nonreciprocal effects in magnetoplasmonic systems supporting simultaneously localized and propagating plasmons, Opt. Express (submitted)

Acknowledgments

Writing a thesis takes a lot of effort and time, sometimes it brings frustration, and sometimes happiness, but at the end, it is worth the pain. Of course, this thesis could not have been completed without the help of countless people, both from the job environment and from outside. People that I recently met, people that I don't see very often, or even people that are there from a long time ago—for all of them, … THANK YOU, and …

lets hope I don't forget anyone!!!

I want to first thank Antonio, my thesis supervisor, for all his help and availability at every moment, no matter the simpleness of the question or the number of times that you have already discussed it before. Furthermore, the joy that he always inspires has been essential for the development of this work. I will follow with Maria, my thesis supervisor; thanks to her a little monster enjoys my everyday. Her dedication in everything she does is complete, what has helped me to learn plenty of things, not all related to magnetoplasmonics. She always helps in everything she can, and she is never bad-tempered. For all of this and a lot more, part of this thesis is also hers.

Lets follow with other group members—I want to thank all "bosses" like Gaspar, Alfonso, and Chemi, the great treatment that I have always received, all the things I have learned with them, and the laughs that I have had at meetings. To my colleagues, what can I say, they are so charming and wonderful, they have helped me a lot, and we have spent great times together!! To those that I met at the beginning: Juan, Elías, Maki, Rui… and to those that came later: Blanca, David, Juan Carlos, César, Huayu, Andreas … as well as to those that have been there all the time: Patri, y Alan, mil gracias, chicos!!

But not all colleagues are magnetoplasmonics!! there are also office colleagues to share morning music, plants … such as Alberto, or those to which you speak shouting to avoid getting up the chair like Jesús, Marina, Mariana, Jaime Andrés, Marta, and o Carmen; or the paddle table players (they will soon published the patent) such as Joselo, Miguel, and Mokri o André. There are also breakfast colleagues, those that fix the world with you, and share thesis or infant traumas!

Carmen, Mercedes, Lorena, Raquel, Estela, Rosa, and Iván P.G. There are upper floor colleagues, those to which you visit when you go up, and get updated: Olga, Sheila, Jaime, Luis Enrique, David Solís, Mario, Alicia, Cris, and y Etor. And there are also Churrasquita colleagues, with whom we go to have lunch at each birthday, or "Jarras" colleagues after a full IMM day! You are all amazing, each of you has contributed to this thesis, thanks! Bego, Elena, Merce, Fran, Horacio, Ryu, Fuster, Malvar, Dani Ramos, Dani Soria, Jero, Valerio, Lukas, Martínez, Marcos, Edu, Piñera, and Villalobos. Those who I met at the beginning have also been relevant, such as Diego, Pablo, Javi, Chon, Z.P., Luisja, Sonia, Nuria, Carlos, and Marian.

Finally, I don't want to forget all IMM staff that make everything possible such as Pulpón, Antonio Lerma, Manolo, Manuel (don't lose your smile!), Mercedes, Margarita, Toña, Nati, Mari Carmen, Carmen, Judith, and y Jota.

I want to acknowledge my foreign collaborators, Vasily and Lionel, from whom I have learned many tricks, and with whom i spent a great time during our meetings; to Rudy, who always writes to me fast and honest! without never having met each other; and finally to Renata, for fruitful scientific discussions and better conversations.

But not everything is thesis-related! That's why I want also to thank …To the beads and other vices girls, laughing with you is a pleasure! To my all the time friends, although we don't see often, we still support each other unconditionally!! To the "Dipolos" that are like my family geek! I met them at University, but I now share more things than that!

I want also to thank … yes, to my family that has always been, and will be there, supporting everything I do, offering me "buenas vibraciones"; it is a joy having you close!!

And last, precisely for being the most important from all, I want to thank you, my "new family." Because you bring me all the energy that I need, because you put at with me, because you turn a bad day into a gorgeous one, and because you, with that little monster that we are raising, are the seed of my happiness

Muchas Zenkius,

and now … read the thesis, the Acknowledgements sections is not enough!!

I want to acknowledge the financial help of the International Iberian Nanotechnology Laboratory, as well as the different projects in which I have take part:

- Optical microsystems Resonant Sensors (MICROSERES) from the Comunidad de Madrid
- Magnetoplasmonics: hybrid nanostructures with magnetic and plasmonic properties from MICINN
- NANOstructured active MAGneto-plasmonic MAterials—NANOMAGMA (NMP3-SL-2008-214107) from EU
- Magneto-optically Active Plasmonic Systems (MAPS) from MICINN

Contents

Abbreviations, Acronyms, and Symbols

L_{sp}	Propagation distance of the surface plasmon polariton
Q	Parameter that relates the magneto-optical constants with the optical ones: $Q = i\frac{\varepsilon_{ij}}{\varepsilon_{ii}}$ for a material
$\Delta^m k_x$	Variation (modulation) of the spp wavevector due to a transverse magnetic field applied
$\Delta^n(\Delta^m k_x)$	Variation of the modulation of the SPP wavevector due to the changes in the refractive index of the dielectric material
$\Delta^n k_x$	Variation of the SPP wavevector due to the changes in the refractive index of the dielectric material
Δk_x	Magnetic modulation of the SPP wavevector
Δk_x^i	Imaginary part of the magnetic modulation of the SPP wavevector
Δk_x^r	Real part of the magnetic modulation of the SPP wavevector
δ	Penetration (vertical) distance of the surface plasmon
ε	Dielectric or optical constant
ε_{ii}	Dielectric or optical constant
ε_{ij}	ij component of the dielectric tensor, i.e., magneto-optical constant
k_0	Wavevector of the light in vacuum
k_x	In-plane component of the SPP wavevector
k_x^i	Imaginary part of the in-plane component of the SPP wavevector
k_x^r	Real part of the in-plane component of the SPP wavevector
k_z	Vertical component of the SPP wavevector
k_z^i	Imaginary part of the vertical component of the SPP wavevector
k_z^r	Real part of the vertical component of the SPP wavevector
n	Refractive index
AC	Alternating signal
AFM	Atomic force microscope
ATR	Attenuated total internal reflection
DC	Continuous signal
DF	Dark field
e-beam	Electron beam

FIB	Focused ion beam
FOM	Figure of merit
IMI	Insulator/metal/insulator; structure consisting of a thin metal layer between two dielectrics
IMM	Instituto de Microelectrónica de Madrid
LSP	Localized surface plasmon
MIM	Metal/insulator/metal; structure consisting of a thin dielectric layer between two metals
MO	Magneto-optic
MP	Magnetoplasmonic
NF	Near field
PGW	Plasmonic gap waveguide
PMMA	Polymethyl methacrylate
PMOKE	Polar magneto-optical Kerr effect
RLC	Electrical circuit consisting of a resistor, an inductor, and a capacitor
SERS	Surface-enhanced Raman scattering
SNOM	Scanning near-field optical microscopy
SP	Surface plasmon
SPP	Surface plasmon polariton
SPR	Surface plasmon resonance
STM	Scanning tunneling microscope
TM	Transverse magnetic
TMOKE	Transverse magneto-optical Kerr effect

Summary

Aim of this thesis

Optics, or photonics, studies the generation, control, and detection of light, as well as light–matter interactions. It is a topic with plenty of applications in many different areas, from telecommunications and computing (signal transport and fast processing), to medicine (laser treatment), sensing (there are optic earthquake sensors), or even the commonly used bar code reading present in every supermarket. Therefore, it is considered as an important research area in relevant funding programs all over the world.

Within photonics, plasmonics focuses in analyzing SPs, which are decaying electromagnetic waves confined in a metal–dielectric interface. SPs possess interesting properties that make them suitable for many possible applications. Due to their evanescent nature and their sensitivity to the materials of then interface, they can be used in sensing, being the most used and analyzed plasmon-based sensors those denoted SPR sensors, which have been proved to be quite sensitive. SPs are also known to provide a strong enhancement of the electromagnetic field at the interface, which has led to the development of surface-enhanced Raman scattering (SERS). Moreover, they can be strongly confined, even beyond the diffraction limit, which can be achieved by nanostructuring the interface surface or using nanoscaled particles. This leads to the development of plasmonic circuitry, which is a key candidate as an alternative to electronic circuitry and traditional optical telecommunication devices, since it is faster than the first one and less bulky than the second. Nowadays, many passive plasmonic devices have been demonstrated, such as waveguides of different lengths and shapes. However, for circuitry, transporting information is not enough, we need to be able to process it. For this reason, one of the targets for plasmonic circuitry is to develop "active" components such as modulators or switches [1–3]. Active implies being able to control their response by means of an external agent such as voltage and temperature. Actually, the main objective of this thesis is to study the effect of an external magnetic field on propagating SPs, also called surface plasmon polaritons (SPPs), for developing active devices.

This thesis is a result of further exploring the potential of "magnetoplasmonics," which combines plasmons with magnetism. Magnetoplasmonics can be used to increase the magneto-optical response of a material due to the local electromagnetic field enhancement associated with the excitation of a plasmon resonance. But it can also be applied in the reverse way, since a magnetic field affects the properties of SPPs, in particular its wavevector. In order to obtain a reasonable magnetic modulation of the SP wavevector, a smart combination of materials that combines suitable ferromagnetic and plasmonic properties, such as a Au/Co/Au trilayer, is required. Once the idea and the materials were decided, we needed a device to measure the propagating SP modulation, which could be done by using a plasmonic interferometer. The idea of applying magnetic field to a plasmonic interferometer (magnetoplasmonic interferometer) was then recently demonstrated using noble/ferromagnetic/noble metallic multilayers, and this thesis is a natural continuation of this initial demonstration. We have used the interferometers as a tool for measuring some the magnetic modulation of the plasmon wavevector, but we have also analyzed its suitability as a device itself, i.e., as a modulator, trying to optimize it and understanding all the processes that take place there. Furthermore, given the potential of plasmonics for sensing, and the accuracy of interferometry, we wanted to evaluate the MP interferometer as a sensor. It was already known that the application of a magnetic field in SPR sensors led to an improvement in their sensitivity; thus, the comparison of SPR sensors with our MP interferometers was a straightforward idea. Finally, due to the local nature of SPs, we have also analyzed the magnetic modulation in the near field by means of some selected interferometric configurations.

The main objectives of this thesis are as follows:

- To understand the underlying physics of applying a magnetic field on propagating surface plasmons using a noble/ferromagnetic/noble metallic structure.
- To take advantage of this effect to implement it in an "active" device, in particular in a plasmonic interferometer that acts as a modulator.
- To study the dependence of the magnetic modulation of propagating SPs on the different parameters, such as the trilayer structure or the magneto-optical material, as well as its spectral behavior and the possibilities to enhance the mentioned magnetic modulation.
- To explore the capability of our magnetically modulated interferometers as sensors, and compare them with the broadly used SPR-based techniques.
- To further extend the understanding of magnetic modulation of surface plasmon polaritons by means of interferometric configurations in the near-field regime.

Results

The first part of the thesis deals with the study of the spectral evolution of the modulation of the SPP wavevector in the visible and near infrared (from 500 nm to 1 μm) for a Au/Co/Au-multilayered system and its optimization. By means of the MP interferometer, we have measured the magnetic modulation of the real part of the SPP wavevector, and of its imaginary part, which had not been done before. It has been seen that the modulation of the real part of the SPP wavevector is larger than the imaginary one for most of the analyzed spectral range. However, for larger wavelengths ($\lambda > 800$), the imaginary part becomes more relevant and should be taken into account to make an accurate description of the system. Regarding the spectral evolution, the SPP wavevector modulation (real and imaginary parts) shows a decreasing behavior with an increasing wavelength. The real part of the modulation Δk_x shows a peak at low wavelengths. By using an analytical expression for Δk_x, we have determined the origin of this spectral evolution and the parameters that determine it. We have established that the spectral trend is due to the evolution of the SPP properties with wavelength and not to the magneto-optical constants. This comes from the vertical confinement of the SPP field and the associated field enhancement inside the Co layer, which can be qualitatively described by a simple parameter: the evolution of the separation between the SPP wavevector and the light line. Considering the use of the magnetoplasmonic interferometers as modulators, we do not measure directly the wavevector magnetic modulation Δk_x, but the transmitted intensity. Therefore, a figure of merit combining both the magnetic modulation and the propagation distance of the SPP ($2\Delta k_x L_{sp}$) is defined and analyzed. In terms of spectral dependence, the decrease in SPP wavevector modulation with the wavelength is overcompensated by the increase in L_{sp} for a significant wavelength range. As a consequence, the 700-nm to 1-μm interval becomes the optimal one for applications.

We have extended this analysis to other ferromagnetic metals such as Fe or Ni, obtaining that the parameters that govern the spectral dependence are again those related to the SPP properties. The figure of merit, on the other hand, is larger when using Fe instead of Co, although we would need larger magnetic fields to saturate the sample.

Still trying to increase the magnetic modulation of the SPP wavevector, we have demonstrated that the deposition of a dielectric overlayer on top of noble/ferromagnetic metal multilayers leads to a significant enhancement of the magnetic field-induced modulation of the SPP wavevector (a 4.5 times increase for an interferometer with a 60-nm PMMA overlayer and a Au top layer of 5−15 nm at 633 nm). The analysis of the figure of merit shows that the modulation depth of a magnetoplasmonic switch can be increased despite of the strong reduction in the SPP propagation length when adding a dielectric layer on top of the metal. This allows to reduce the size of the device, and therefore, this finding represents an essential step toward miniaturization of active magnetoplasmonic devices.

Finally, from the analysis of the figure of merit, we have determined that a magnetoplasmonic modulator consisting of a Au/Fe/Au interferometer with a separation distance of $3L_{sp}$ and covered with a dielectric with $n_d = 1.49$ could provide intensity modulations of around 12 % in the optimal spectral range (950 nm). This value is not so far from other integrated plasmonic modulator performances reported in the literature based on the thermo-optical or electro-optical effects.

The second part of the thesis is related to the theoretical analysis of the capability of plasmonic (without magnetic field) and magnetoplasmonic interferometers as biosensors. A comparison of the sensibility of a plasmonic interferometer and a SPR sensor is made. Also, the sensitivity of the two interferometers is compared. In order to perform a fair comparison, we have theoretically optimized the thickness of the metallic layers for each configuration and wavelength using the same noble metal, Au. We have compared the sensitivity of the three methods to changes in the refractive index of the sensing layer. However, as each method measures a different magnitude, i.e., measures a different output, both the sensitivity of the physical parameter implied in the measurement and that of the output have been analyzed. Both SPR and plasmonic interferometry rely on the modification of the SPP wavevector under refractive index variations, $k_x(n)$. Being the physical system supporting each of them very similar, Au films, the sensitivity of the wavevector $\left(\frac{\delta k_x}{\delta n}\right)$ is very similar for both of them. When analyzing the measured output, it has been shown that the plasmonic interferometer can surpass the SPR in sensitivity. This is due to the fact that the sensitivity of the plasmonic interferometer is proportional to the slit–groove distance, so by increasing this distance the sensitivity can be enhanced. Even though the slit–groove distance is enlarged to improve the plasmonic interferometer sensitivity, it is still kept on the order of a few tens of microns, which allows its integration in a chip for the development of miniaturized sensors. For the magnetoplasmonic interferometer, two physical parameters are involved in a sensing experiment: the modification of the SPP wavevector $k_x(n)$ and the variation of the magnetic field-driven SPP wavevector modulation $\Delta^m k_x(n)$. This last quantity has a stronger dependence with the refractive index than $k_x(n)$, so a sensor based on it would provide a higher performance. However, since $\Delta^m k_x$ is several orders of magnitude smaller than k_x, its stronger sensitivity can only be appreciated if the monitored output quantity in a magnetoplasmonic interferometer is directly related to $\Delta^m k_x$, without the k_x contribution.

Regarding the spectral dependence of the sensitivity, this increases with the wavelength for both the SPR method and the plasmonic interferometer, even though the opposite behavior of the sensitivity of $k_x(n)$ $\left(\frac{\delta k_x}{\delta n}\right)$ suggested otherwise. This is due to the fact that for large wavelengths, the SPP has less losses; thus, the SPR resonances are narrower and the propagation distance L_{sp} is larger, allowing larger distances between the groove and the slit for the interferometer. On the other hand, for an ideal magnetoplasmonic interferometer (i.e., only dependent on $\Delta^m k_x$), we should work at low wavelengths, since the sensitivity of $\Delta^m k_x$ $\left(\frac{\delta \Delta^m k_x}{\delta n}\right)$ decays with the wavelength.

The last part of the manuscript shows the possibilities of magnetic modulation in different near-field interferometric configurations. The magnetic field-induced modulation of k_x could be detected in the near field by measuring the interference of two counterpropagating plasmons between two launching slits. The near-field intensity modulation would be proportional to $4\Delta k_x \times d$, being $2d$ the distance between the two slits. It could also be possible to have direct access to the magnetic modulation of k_x^i and even of k_z by measuring the magnetic modulation of a decaying SPP in an isolated slit.

Conclusions

Five essential conclusions can be extracted from this work:

- Magnetoplasmonic interferometry is a very interesting tool to analyze the magnetic modulation of the SPP wavevector.
- The parameters that govern the modulation depth of the magnetoplasmonic interferometers are mainly related to the properties of the SPPs, rather than to the magneto-optical properties of the ferromagnetic material. It is specially relevant to the role of the electromagnetic field distribution along the materials of the interface, as well as the dispersion relation of the SPPs and how far or close it is from the line of the light, $(k_x - k_0)$.
- A magnetoplasmonic interferometer can be a device per se. It is a reasonable choice to modulate SPPs. Applying external magnetic fields of 20 mT, we have been able to obtain modulation depths of around 2 % on non-optimized Au/Co/Au trilayers, with interferometer sizes on the microscale. Moreover, we could obtain modulation depths of about 12 % by substituting Co by Fe, working at around 950 nm and covering the interferometer with a dielectric overlayer of $n_d = 1.49$.
- Plasmonic interferometry is interesting also for biosensing, where it can be more sensitive than traditional SPR sensors. Magnetoplasmonic interferometry offers a new sensing parameter, the magnetic modulation of the SPP, which is more sensitive to refractive index variations.
- The magnetic modulation of SPP could also be obtained from near-field experiments. With the appropriate configurations, we could obtain not only the modulation of the SPP wavevector k_x, but also that of its vertical component k_z.

References

1. E. Ozbay, Science **311**, 189–193 (2006)
2. A.V. Krasavin, N.I. Zheludev, Appl. Phys. Lett. **84**, 1416–1418 (2004)
3. K.F. MacDonald, N.I. Zheludev, Laser Photon. Rev. **4**, 562 (2010)

Chapter 1
Preamble

Optics, or photonics, studies the generation, control and detection of light, as well as light-matter interactions. It is a topic with plenty of applications in many different areas, from telecommunications and computing (signal transport and fast processing), to medicine (laser treatment), sensing (there are optic earthquake sensors), or even the commonly used bar code reading present in every supermarket. It is considered as an important research area in relevant funding programs both in USA [1] and in Europe, where €47 million have been budgeted to different photonic areas, including plasmonics [2–4].

Within photonics, plasmonics [5, 6] is interested in analyzing SPs, which are decaying electromagnetic waves confined in a metal-dielectric interface. Their evanescent nature and their sensitivity to the materials of the interface make them suitable for many possible applications such as sensing, being the most used and analyzed plasmon-based sensors those denoted surface plasmon resonance (SPR) sensors, which have been proved to be quite sensitive [7]. SPs are also known to provide a strong enhancement of the electromagnetic field at the interface, which has lead to one of their most popular applications: Surface Enhanced Raman Spectroscopy [8] (SERS). Moreover, they can be strongly confined, even beyond the diffraction limit, which can be achieved by nanostructuring the interface surface or using nanoscaled particles [9, 10]. This leads to other potential applications, such as plasmon-based optical trapping [11] or plasmonic circuitry [12], where plasmonics is a key candidate as an alternative to electronic circuitry and traditional optical telecommunication devices, being faster than the first one and less bulky than the second. Nowadays, many passive plasmonic devices have been demonstrated, such as waveguides of different lengths and shapes. However, for circuitry, transporting information is not enough, we also need to be able to process it. For this reason, one of the targets for plasmonic circuitry is to develop "active" components such as modulators or switches [12–14]. This implies being able to control the SPs response by means of an external agent such as voltage, temperature... Actually, the main objective of this thesis is to study the effect of an external magnetic field on propagating SPs for developing active devices.

D. Martín Becerra, *Active Plasmonic Devices*, Springer Theses,
DOI 10.1007/978-3-319-48411-2_1

This thesis is a result of further exploring the potential of "magnetoplasmonics", which combines plasmons with magnetism. With this concept as the main area of interest, the Magnetoplasmonic Nanostructures group at the Instituto de Microelectrónica de Madrid was born in 2006. Traditionally (and also within the group), magnetoplasmonics was first applied to enhance the magnetooptical response of a material by the local electromagnetic field enhancement due to the excitation of a plasmon [15–18]. Soon it was seen that this could also work the other way around, and that a magnetic field could affect the propagating SP properties [19], leading to active plasmonics. Moreover, in order to obtain a reasonable magnetic modulation of the plasmon properties, a smart combination of materials that combines suitable ferromagnetic and plasmonic properties, such as a Au/Co/Au trilayer is required. Once the idea and the materials were decided, a device to measure the propagating SP modulation was needed, and there were in the literature some "plasmonic" interferometers used in different configurations [20–22] for that task. The idea of applying a magnetic field to a plasmonic interferometer (magnetoplasmonic interferometer) was then recently demonstrated using noble/ferromagnetic/noble metallic multilayers [23].

This thesis is a natural continuation on the initial demonstration shown in Ref. [23]. We have used these magnetoplasmonic interferometers as a tool for measuring the magnetic modulation of the plasmon wavevector, but we have also analyzed its suitability as a device itself, i.e., as a modulator, trying to optimize it and to understand all the processes that take place there. Furthermore, given the potential of plasmonics for sensing and the accuracy of interferometry, we have evaluated the magnetoplasmonic (MP) interferometer as a sensor. It was already known that the application of a magnetic field in SPR sensors led to an improvement in their sensitivity [24, 25]; thus the comparison of SPR sensors with our MP interferometers was a straightforward idea. Finally, due to the local nature of SPs, we are going to analyze the effect of applying a magnetic field on propagating SP in the near field regime. For that, some alternative interferometric plasmonic configurations proposed in the literature for near field studies [26, 27] have been considered.
The main objectives of this thesis are, therefore:

- To understand the underlying physics of applying a magnetic field on propagating surface plasmons using a noble/ferromagnetic/noble metallic structure.
- To take advantage of this effect to implement it in an "active" device, in particular in a plasmonic interferometer that acts as a modulator.
- To study the dependence of the magnetic modulation of propagating SPs on the different parameters involved in the interferometers, such as the trilayer structure or the magnetic material, as well as their spectral behavior and the possibilities to enhance the mentioned magnetic modulation.
- To explore the capability of our magnetically modulated interferometers as sensors, and compare them with the broadly used SPR based techniques.
- To further extend the understanding of magnetic modulation of surface plasmon polaritons by means of interferometric configurations in the near field regime.

The first three objectives have been addressed experimentally and with the support of numerical simulations. Based on the understanding obtained from this, we have approached the last two numerically.

References

1. www.osa.org/en-us/about_osa/public_policy/national_photonics_initiative/harnessing_light/
2. www.photonics.com/Article.aspx?AID=55531
3. www.photonics21.org/
4. http://ec.europa.eu/programmes/horizon2020/en
5. S.A. Maier, *Plasmonics: Fundamentals and Applications* (Springer, Berlin, 2007)
6. H. Raether, *Surface Plasmons* (Springer, Berlin, 1986)
7. J. Homola, S.S. Yee, G. Gauglitz, Sens. Actuators B **54**, 3–15 (1999)
8. K.A. Willets, R.P. Van Duyne, Annu. Rev. Phys. Chem. **58**, 267–297 (2007)
9. R. Zia, J.A. Schuller, A. Chandran, M.L. Brongersma, Adv. Mater. **9**, 20–27 (2006)
10. W.L. Barnes, A. Dereux, T.W. Ebbesen, Nature **424**, 824–830 (2003)
11. M.L. Juan, M. Righini, R. Quidant, Nat. Photonics **5**, 349–356 (2011)
12. E. Ozbay, Science **311**, 189–193 (2006)
13. A.V. Krasavin, N.I. Zheludev, Appl. Phys. Lett. **84**, 1416–1418 (2004)
14. K.F. MacDonald, N.I. Zheludev, Laser Photonics Rev. **4**, 562 (2010)
15. J.B. Gonzalez-Diaz, A. Garcia-Martin, J.M. Garcia-Martin, A. Cebollada, G. Armelles, B. Sepulveda, Y. Alaverdyan, M. Kll, Small **4**, 202–205 (2008)
16. J.B. Gonzalez-Diaz, J.M. Garcia-Martin, A. Garcia-Martin, D. Navas, A. Asenjo, M. Vazquez, M. Hernandez-Velez, G. Armelles, Appl. Phys. Lett. **94**, 263101 (2009)
17. E. Ferreiro-Vila, M. Iglesias, E. Paz, F.J. Palomares, F. Cebollada, J.M. Gonzalez, G. Armelles, J.M. Garcia-Martin, A. Cebollada, Phys. Rev. B **83**, 205120 (2011)
18. G. Armelles, A. Cebollada, A. Garcia-Martin, M.U. Gonzalez, Adv. Opt. Mater. **1**, 10–35 (2013)
19. J.B. Gonzalez-Diaz, A. Garcia-Martin, G. Armelles, J.M. Garcia-Martin, C. Clavero, A. Cebollada, R.A. Lukaszew, J.R. Skuza, D.P. Kumah, R. Clarke, Phys. Rev. B **76**, 153402 (2007)
20. G. Gay, O. Alloschery, B. Viaris de Lesegno, C. O'Dwyer, J. Weiner, H.J. Lezec, Nature **2**, 262–267 (2006)
21. V.V. Temnov, U. Woggon, J. Dintinger, E. Devaux, T.W. Ebbesen, Opt. Lett. **32**, 1235–1237 (2007)
22. V.V. Temnov, K. Nelson, G. Armelles, A. Cebollada, T. Thomay, A. Leitenstorfer, R. Bratschitsch, Opt. Express **17**, 8423–8432 (2009)
23. V.V. Temnov, G. Armelles, U. Woggon, D. Guzatov, A. Cebollada, A. Garcia-Martin, J.M. Garcia-Martin, T. Thomay, A. Leitenstorfer, R. Bratschitsch, Nat. Photonics **4**, 107–111 (2010)
24. D. Regatos, D. Farina, A. Calle, A. Cebollada, B. Sepulveda, G. Armelles, L.M. Lechuga, J. Appl. Phys. **108**, 054502 (2010)
25. M.G. Manera, G. Montagna, E. Ferreiro-Vila, L. Gonzalez-Garcia, J.R. Sanchez-Valencia, A.R. Gonzalez-Elipe, A. Cebollada, J.M. Garcia-Martin, A. Garcia-Martin, G. Armelles, R. Rella, J. Mater. Chem. **21**, 16049–16056 (2011)
26. B. Wang, L. Aigouy, E. Bourhis, J. Gierak, J.P. Hugonin, P. Lalanne, Appl. Phys. Lett. **94**, 011114 (2009)
27. L. Aigouy, P. Lalanne, J.P. Hugonin, G. Julié, V. Mathet, M. Mortier, Phys. Rev. Lett. **98**, 153902 (2007)

Chapter 2
Introduction to Active Plasmonics and Magnetoplasmonics

This chapter includes the basic knowledge to understand this thesis. Concepts such as SPP, i.e., propagating SPs; with the corresponding dispersion relation and distribution of the electromagnetic field inside the materials, are explained. In addition, the idea of active plasmonics and magnetoplasmonics, together with the importance of the used materials, is presented. Moreover, the modulation of the SPP wavevector by means of a magnetic field (which is a key point in this manuscript) is shown.

2.1 Surface Plasmons

The ability to obtain intense colors and different illumination effects by incorporating small amounts of noble metals into glass was already known since the late roman age, as in the famous Lycurgus cup (4th century AD[1]) and in the stained glasses of medieval churches or cathedrals. However, the physical explanation of this phenomenon arrived much later, at the beginning of the 20th century. So, Garnett [2] in 1904 used the recently developed Drude theory of the metals to explain the bright colors observed in metal doped glasses; and Mie in 1908 produced his, now well known, theory of light scattering by particles [3]. In this way, the bright coloring was associated with the presence of noble metal nanoparticles and the excitation of LSP.

Surface Plasmons are oscillations of the free electrons in the metal induced by electromagnetic radiation, and are bound to the interface between a metal and a dielectric. They appear as a peak (resonance) in the optical response [4] and are characterized by an enhancement of the electromagnetic field at this interface. They can be localized (LSP) [5] when they take place at closed interfaces such as nanostructures, or they can be propagative [6] when they take place at flat interfaces [4, 7].

The mathematical description for the propagating surface plasmons, called SPPs, is known since around 1900, when Sommerfeld and Zenneck discovered

[1] This cup is nowadays shown at the Brithish Museum [1].

D. Martín Becerra, *Active Plasmonic Devices*, Springer Theses,
DOI 10.1007/978-3-319-48411-2_2

electromagnetic waves at a metal surface at the radio frequencies [8, 9]. At the same time, Wood observed unexplained intensity drops in the spectra of the reflection of light in metallic gratings for visible frequencies [10], which became known as Wood anomalies. Some years later, Fano, Pines, and Ritchie [11–13] connected those Wood Anomalies to SPP resonances. Finally, Otto, Krestchmann and Raether proposed a method to excite the surface plasma waves with visible light by using a prism [14, 15] and a unified and complete description of all these phenomena in the form of Surface Plasmon Polaritons was established [4].

Although surface plasmons began to be known at the beginning of the 20th century, and finally understood around the 70s, they have attracted a renewed interest over the last decades due to their properties and the access to the nanometer scale that has been acquired with the development of nanofabrication tools. Indeed, the confinement of the plasmon electromagnetic field within volumes of the order of or smaller than the wavelength, their local electromagnetic field enhancement and the dependence of their properties on the materials forming the interface makes them suitable for an endless list of applications, including photothermal therapy [16], $3D$ holography [17], or quantum information processing [18]. Traditionally, two of the most used applications are their use for single molecule detection using SERS [19–21], and the extraordinary transmission through periodic hole arrays [22, 23]. Besides, due to the confinement of the electromagnetic field of the plasmon at the interface, they are very sensitive to any change in the refractive index of the materials of this interface and diverse sensing configurations have been successfully proposed, with both propagating [24] and localized plasmons [25]. Finally, plasmons are also interesting for the development of nanophotonic circuitry [26–30], where SP could be fundamental to replace traditional electronic (small but slow) or optical-fiber (fast but large) circuitry in order to get fast and small devices. The contributions of this thesis lie within the scope of these two last applications.

2.1.1 Surface Plasmon Polaritons

As mentioned above, we want to contribute to the further development of surface plasmon based devices, not only to finally achieve plasmonic circuits but also for sensing. In both cases we have worked with propagating surface plasmon polaritons (SPPs), whose properties I will describe in detail in the following.

SPPs are exponentially decaying transverse magnetic (TM) electromagnetic waves confined to a flat metal-dielectric interface, and are related to free electron oscillations [4]. To obtain the characteristics of SPPs, we can consider a single interface between two semi-infinite media with dielectric constants ε_d and ε_m in the XY plane, as sketched in Fig. 2.1. The wave equation that comes from Maxwell equations can be expressed as:

$$\frac{\partial^2 \vec{E}(z)}{\partial z^2} + ({k_0}^2\varepsilon - \Omega^2)\vec{E} = 0, \tag{2.1}$$

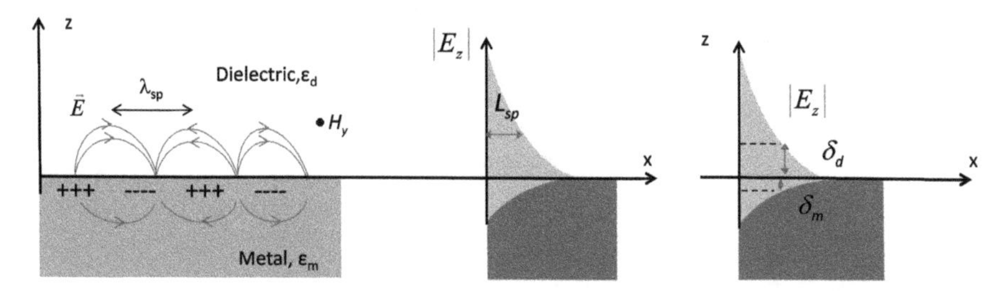

Fig. 2.1 Schematic representation of a SPP. Four important parameters of the SPP are shown: its wavelength, related to the SPP wavevector by $\lambda_{sp} = \frac{2\pi}{k_x}$; the propagation distance L_{sp}; and the penetration distances (δ) on both the dielectric and the metal, δ_d and δ_m

being k_0 the wavevector of light in vacuum, Ω the propagative wavevector of the wave, ε the dielectric permittivity of the analyzed material (ε_m or ε_d depending on the side of the interface), and $\vec{E}$ the electric field of the wave that only depends on z due to the geometry of the interface. An equivalent expression exists for the magnetic field $\vec{H}$. Within the solutions of this expression we are looking for a wave with the SPP characteristics; i.e., an evanescent wave that propagates along the x direction while z is the vertical direction, being $z > 0$ the dielectric side and $z < 0$ the metallic one:

$$\begin{aligned} \vec{E} &= \vec{E}_0^d \cdot e^{ik_x x} e^{ik_{z,d} z} e^{-i\omega t} \quad \text{for } z > 0, \\ \vec{E} &= \vec{E}_0^m \cdot e^{ik_x x} e^{-ik_{z,m} z} e^{-i\omega t} \quad \text{for } z < 0, \end{aligned} \tag{2.2}$$

where k_x ($k_x = \Omega$) and k_z are the in plane and vertical component of the SPP wavevector, $E_0^{d,m}$ are the amplitudes of the wave at each side of the interface, and ωt represents the temporal dependence. Furthermore, k_z is purely imaginary[2] since we are considering an evanescent wave, and it is different for the dielectric ($k_{z,d}$) and for the metal ($k_{z,m}$), meaning that the electromagnetic field of the SPP will penetrate differently into each material of the interface. The electromagnetic fields are related through Maxwell equations [4, 7], and for a TM mode they can we written as:

$$\begin{aligned} H_y &= -\frac{\omega \varepsilon_0 \varepsilon}{k_x} E_z, \\ E_x &= -i \frac{1}{k_x} \frac{\partial E_z}{\partial z}, \end{aligned} \tag{2.3}$$

where ε_0 is the dielectric permittivity in vacuum, E_x and E_z are the x and z components of the electric field of the SPP, and H_y is the y component of its magnetic field. For $z > 0$, taking E_z of the form of Eq. 2.2, applying Eq. 2.3, and integrating the temporal dependence and the amplitude into the constant A^j we obtain that the fields can be expressed as [4, 7]:

[2]For an ideal metal, with no losses.

$$
\begin{aligned}
E_z^d &= A^d e^{i(k_x x + k_{z,d} z)} \\
E_x^d &= A^d \frac{k_{z,\,d}}{k_x} e^{i(k_x x + k_{z,d} z)} \\
H_y^d &= -A^d \frac{\omega \varepsilon_0 \varepsilon_d}{k_x} e^{i(k_x x + k_{z,d} z)},
\end{aligned}
\tag{2.4}
$$

on the other hand, for $z < 0$ the fields are expressed as:

$$
\begin{aligned}
E_z^m &= A^m e^{i(k_x x - k_{z,m} z)} \\
E_x^m &= -A^m \frac{k_{z,\,m}}{k_x} e^{i(k_x x - k_{z,m} z)} \\
H_y^m &= -A^m \frac{\omega \varepsilon_0 \varepsilon_m}{k_x} e^{i(k_x x - k_{z,m} z)}.
\end{aligned}
\tag{2.5}
$$

For a TM wave the wave equation shown before can be written as:

$$
\frac{\partial^2 E_z}{\partial z^2} + ({k_0}^2 \varepsilon - k_x^2) E_z = 0 \tag{2.6}
$$

Using Eqs. 2.4, 2.5, and 2.6 we obtain the expression for the vertical component of the SPP wavevector, k_z, which is complex. It is defined for each material j (and thus for each side of the interface $z > 0$ or $z < 0$) as:

$$
k_{z,j} = i k_{z,j}^i = \sqrt{k_0^2 \varepsilon_j - k_x^2}, \tag{2.7}
$$

where ε_j is the dielectric constant of the material j, and k_x is the in-plane component of the SPP wavevector.

Finally, applying the boundary conditions (at $z = 0$) for this wave; i.e., continuity of $\varepsilon_j E_z$, H_y and E_x we find that:

$$
\frac{k_{z,m}}{k_{z,d}} = -\frac{\varepsilon_m}{\varepsilon_d}. \tag{2.8}
$$

The combination of this last expressions (Eqs. 2.7 and 2.8) for both the metal and the dielectric sides lead to the SPP dispersion relation, which in this particular case of a SPP sustained at the interface between two semi-infinite mediums in the XY plane and propagating along the x direction is [4, 7]:

$$
k_x = k_0 \sqrt{\frac{\varepsilon_m \varepsilon_d}{\varepsilon_m + \varepsilon_d}}. \tag{2.9}
$$

From both Eqs. 2.8 and 2.9 it can be seen that SPP take place at interfaces where $\varepsilon_d > 0$ and $\varepsilon_m < 0$, fulfilling that $\varepsilon_d + \varepsilon_m \neq 0$. Therefore, surface plasmons can only exist for interfaces of materials with dielectric constants of opposite signs, and in general for metal-dielectric. Note that, although the SPP wavevector k_{sp} consists of two components k_x and k_z, usually k_{sp} and k_x are used indistinctly, and both ways can be found in this thesis. Moreover, normally $k_z \ll k_x$, then in those cases, we can assume that $E \approx E_z$, and the SPP electric field for the dielectric and the metal can be expressed as:

$$
\begin{aligned}
E \approx E_z &= A^d e^{ik_x x} e^{-k^i_{z,d} z}, \text{ for } z > 0 \\
E \approx E_z &= A^d \frac{\varepsilon_d}{\varepsilon_m} e^{ik_x x} e^{k^i_{z,m} z}, \text{ for } z < 0,
\end{aligned}
\tag{2.10}
$$

where z is positive for the dielectric and negative for the metallic side. The expressions of the SPP electromagnetic field and the dispersion relation contain all the properties of surface plasmon polaritons. Next, two important aspects are considered: the meaning of the SPP parameters and the SPP excitation.

- SPP parameters description

The expressions of the SPP electromagnetic field (Eqs. 2.10) show that it has a strongly localized nature along the interface, that decays exponentially at both sides. Figure 2.1 shows this decay and several fundamental parameters of SPP: the SPP wavelength λ_{sp}, its propagation distance L_{sp}, and the penetration depth of the electromagnetic field in the j material δ_j. These properties are related to the dispersion relation of SPP and to the vertical component of the SPP wavevector k_z, given by Eqs. 2.7 and 2.9.

As it can be seen from Eq. 2.9, k_x depends on the optical properties of the materials of the interface. Besides, the dielectric constants of the metal are complex (metals have absorption), and therefore the wavevector of the SPP will be complex too, $k_x = k^r_x + ik^i_x$. Then, for $z > 0$ for example, Eq. 2.10 can be rewritten as:

$$
E_z = Ae^{ik^r_x x} e^{-k^i_x x} e^{-k^i_{z,d} z}.
\tag{2.11}
$$

The real part k^R_x is related to the phase velocity of the wave, and the imaginary part k^i_x is related to the losses of the wave, and thus to the propagation distance of the plasmon. As it can be seen in the Eq. (2.11), the SPP attenuation in the propagation direction x is related to the absorption of the metal. Assuming that ε_d is real and that $\varepsilon^i_m < |\varepsilon^r_m|$, the real and imaginary parts of k_x can be derived from Eq. 2.9 [7]:

$$
\begin{aligned}
k^r_x &= k_0 \sqrt{\frac{\varepsilon^r_m \varepsilon_d}{\varepsilon^r_m + \varepsilon_d}}, \\
k^i_x &= k_0 \frac{\varepsilon^i_m}{2 \cdot (\varepsilon^r_m)^2} \left(\frac{\varepsilon^r_m \varepsilon_d}{\varepsilon^r_m + \varepsilon_d} \right)^{\frac{3}{2}},
\end{aligned}
\tag{2.12}
$$

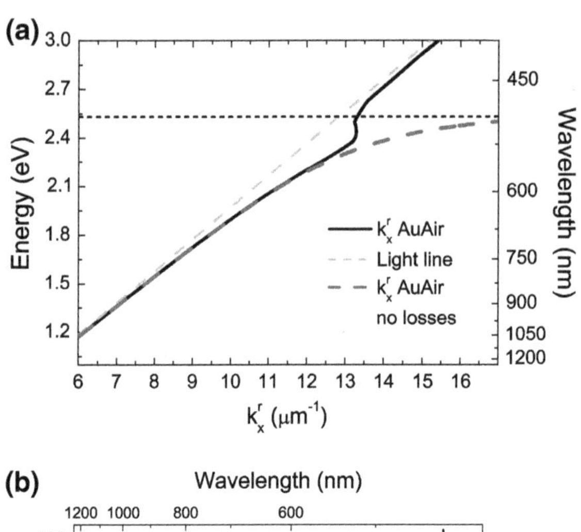

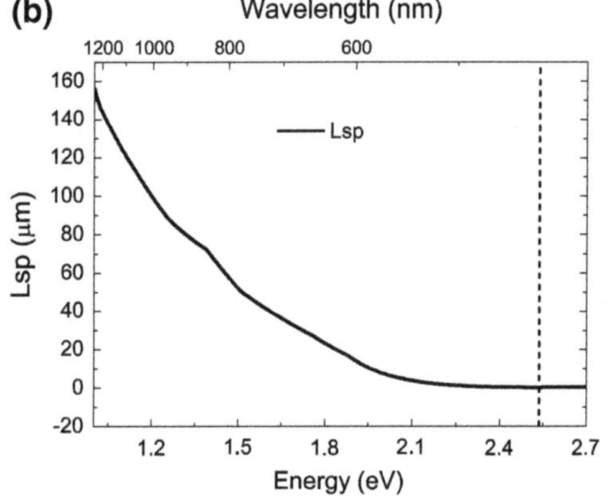

Fig. 2.2 Dispersion relation for a SPP at a Au-air interface. **a** Real part of the SPP wavevector k_x^r (*solid black line*) compared with the ideal situation where there are no losses (*thick red dashed line*), which means there is not imaginary part of the dielectric constant of the metal. As a reference, the light line k_0 is also shown (*grey dashed line*). **b** Propagation distance of the SPP L_{sp}. The *dotted black line* at about 2.6 eV represents the asymptotic value E_p where the SPP basic conditions are not fulfilled anymore

The propagation distance of the SPP is defined as $L_{sp} = 1/(2k_x^i)$, and represents the distance at which the intensity of the SPP electromagnetic field decreases a factor $1/e$. For an ideal metal (with no losses, $\varepsilon_m^i = 0$), there will be no attenuation and the SPP would propagate indefinitely. However, metals always have absorption, and therefore the propagation of the SPP is limited, being the propagation distances of the order of micrometers, depending on the properties of the neighboring materials and on the excitation wavelength. In fact, this intrinsic loss of the system is one of the main drawbacks of SPP for circuits, although it can be compensated by using appropriate gain strategies [31, 32]. Figure 2.2 shows the values of k_x^r and L_{sp} for a SPP propagating at a single Au-Air interface. For these and all the calculations shown in this section, the dielectric constants of Au have been obtained from Johnson and Christy's work [33]. In Fig. 2.2 we have also included the dispersion relation for a metal with no losses (dashed line). In this case, the dispersion relation approaches an asymptotic energy value $E_{sp} = \frac{E_p}{\sqrt{2}}$ in air, where E_p is the energy equivalent to the plasma frequency. This E_{sp} limit value is related to the situation where $\left|\varepsilon_m^r\right| \cong \varepsilon_d$, and the basic condition to have SPP is not fulfilled [4, 7] (the asymptote is represented

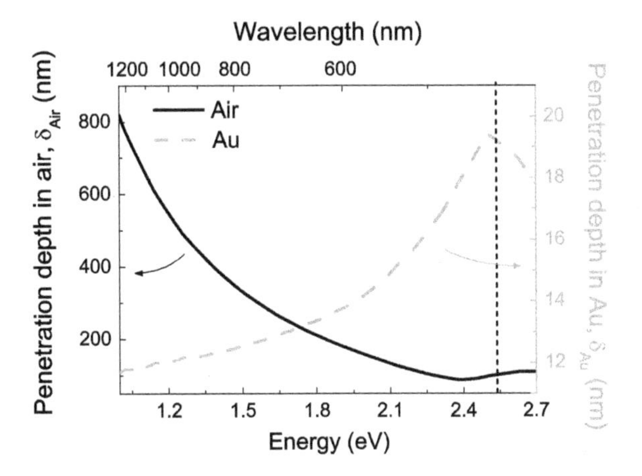

Fig. 2.3 Penetration distance of a SPP at each side of a Au-air interface as a function of the wavelength

in Fig. 2.2 as the black dotted line). When the metal has losses ($\varepsilon_m^i \neq 0$), even though $\left|\varepsilon_m^r\right| \cong \varepsilon_d$ there is not an indetermination in k_x (see Eq. 2.9), so instead of an asymptote, the dispersion relation bends and it can even cross the light line (solid line in Fig. 2.2). However, there the plasmonic conditions are not fulfilled, the SPP "nature" is lost and the wave becomes a radiative or leaky mode. Finally, in Fig. 2.2 it can be seen that for larger energies (always below the dotted curve), the real part of the SPP wavevector gets larger and further deviates from the line of the light (grey dashed line), while there is a decrease in the propagation distance. For smaller energies, on the other hand, k_x^r is smaller and closer to the line of the light, and there is an increase of the propagation distance.

Similarly to the propagation distance, the penetration depth of the electromagnetic field in the j material can be defined as the vertical distance at which the electromagnetic field intensity decays up to a factor $1/e$. As it can be seen from Eq. 2.10, this implies that $\delta_j = 1/(2k_{z,j}^i)$. In Fig. 2.3, this exponential decay for a single Au-air interface is represented. There it can be seen that the penetration of the SPP field is lower in the metallic layer than in the dielectric, as expected, and then the vertical component of the SPP wavevector is larger for the metal than for the dielectric (see Eq. 2.7).

The above described SPP properties (penetration distances δ_j and propagation distance L_{sp} together with k_x^r) represent indeed a way of describing the SPP electromagnetic field and whether it is more or less confined. Actually, all of these parameters are related, by Eqs. 2.7, 2.9, and 2.10. So, looking at the spectral dependence, from Figs. 2.2, 2.3 and 2.4 we can see that for low energies k_x^r is closer to the light line. Therefore k_z^d is smaller and δ_d is larger, hence the associated electromagnetic field is more spread out of the interface, at the dielectric material. Then it is less confined and it propagates longer distances. On the other hand, for larger wavelengths, k_x^r gets further apart from the light line, so k_z^d increases and δ_d decreases, thus the SPP field becomes more confined to the interface and the presence of the metallic layer is more relevant, being the propagation distances smaller. A similar description can be done

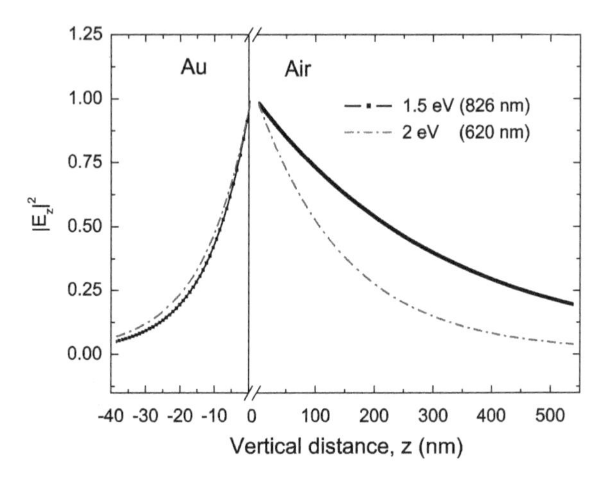

Fig. 2.4 Electromagnetic field of the SPP in a Au-air interface as a function of the vertical distance z for 620 and 826 nm. The left region ($z < 0$) represents the metal (Au), while the right one ($z > 0$) is the dielectric (air). Please note that the scale for the x-axis of the graph is different for $z < 0$ and $z > 0$. The *break* represents the $z = 0$ interface point

by means of the penetration of the SPP electromagnetic field into the metallic side, however, the changes in the penetration distance of the metal are quite smaller than in the case of the dielectric. This is shown in Fig. 2.4, where the distribution of $|E_z|^2$ at both sides of the interface is shown for two different energies.

- SPP excitation

As it has been mentioned before, k_x depends on the properties of the two materials of the interface (Eq. 2.9). Moreover, as $|\varepsilon_m| > |\varepsilon_d|$, this wavevector is always larger than the wavevector of light in the dielectric medium ($k_x > \sqrt{\varepsilon_d} k_0$) or, in other words, the SPP dispersion relation is beyond the light line (as can be seen in the representation of Eq. 2.9 as a function of the energy shown in Fig. 2.2). This implies that the SPP cannot be excited directly impinging with light from the dielectric side of the interface, since the component of the wavevector of the light parallel to the interface ($k_{II} = k_0 \sin\theta$), is always smaller than k_x, and has to be conserved. Indeed, this is why the first SPP analysis were done using fast electron excitation [12, 13]. To overcome this problem, we need to provide to the incident light an extra momentum, which can be done in several ways.

The coupling of the SPP can be achieved using a prism over a thin metallic film. As it is shown in Fig. 2.5, the light impinges through the prism and the SPP is launched at the other interface of the metal [14, 15]. This way, due to the presence of the prism, the parallel component of the wavevector of the incident light is then $k_{II} = n_p k_0 \sin\theta$. If the refractive index of the prism is larger than the one of the dielectric used, $n_p > n_d$, then $k_{II} \geq k_x \approx k_0 \sqrt{(}\varepsilon_d)$, which will lead to SPP excitation at certain angles, exactly those verifying $k_x = n_p k_0 \sin\theta$. However, it is important to notice that, the launched SPP will be the one at the metal-dielectric interface, not the prism-metal SPP (that has a larger k_x), which limits the use of this method only to thin films. Moreover, the angle of incidence to launch the SPP will depend on the wavelength used. This method is called Kretschmann or ATR and provided that is very used for sensing, it will be considered in Chap. 5 of this thesis. Other procedure

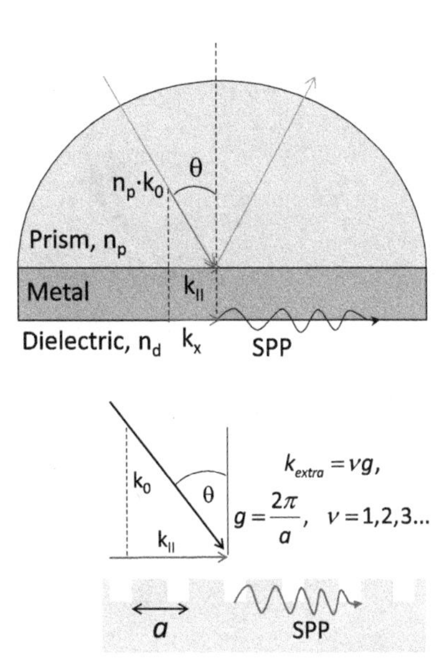

Fig. 2.5 SPP excitation in a thin metallic layer by means of a prism

Fig. 2.6 SPP excitation by means of a grating with lattice constant a and reciprocal vector g

to excite SPPs consists in the use of a grating [26] on the metal surface (Fig. 2.6). This method can be applied to metal layers of any thickness. The extra momentum is provided by the reciprocal vector of the grating, g, resulting in the following SPP wavevector $k_x = k_0 \sin\theta + \nu g$. The angle of incidence and the parameters of the grating will depend on the wavelength at which we want to work. Finally, a small defect [34] on the metal surface can be also used to launch SPPs. This last is in fact the main method used in this thesis for SPP excitation. This case is very similar to the grating situation, the extra momentum is going to be determined by the light scattered from the defect, being the SPP wavevector then $k_x = k_0 \sin\theta + \Delta_{defect}k$, being $\Delta_{defect}k$ all the possible directions and magnitudes of the extra momentum depending on the kind of defect we have. When the incident light has a component of the electric field parallel to the extra momentum supplied by the defect, a SPP will be launched in that direction (Fig. 2.7). For example, a dot will provide an extra momentum (scattered light) in all directions of the interface plane, therefore the SPP launching will be marked by the polarization of the incident light (see Fig. 2.7b, d, f). A line, on the other hand, provides a momentum (scatters light) perpendicular to its longest side, thus the SPP will be excited only when the electric field is in that direction too [4, 7, 34] (Fig. 2.7a, c, e). Using this last method the SPP can be excited for any wavelength. Nevertheless, depending on the size of the defect and on the incident wavelength the SPP excitation will be more or less efficient [35, 36].

Summarizing, we have seen in this basic description of SPPs, using the simplest geometry of a single interface, two important properties that motivate the SPP interest for the applications of concern within this thesis:

- vertical confinement of the electromagnetic field, for miniaturized circuits.
- k_x dependence on the dielectric properties ε_d, for sensing.

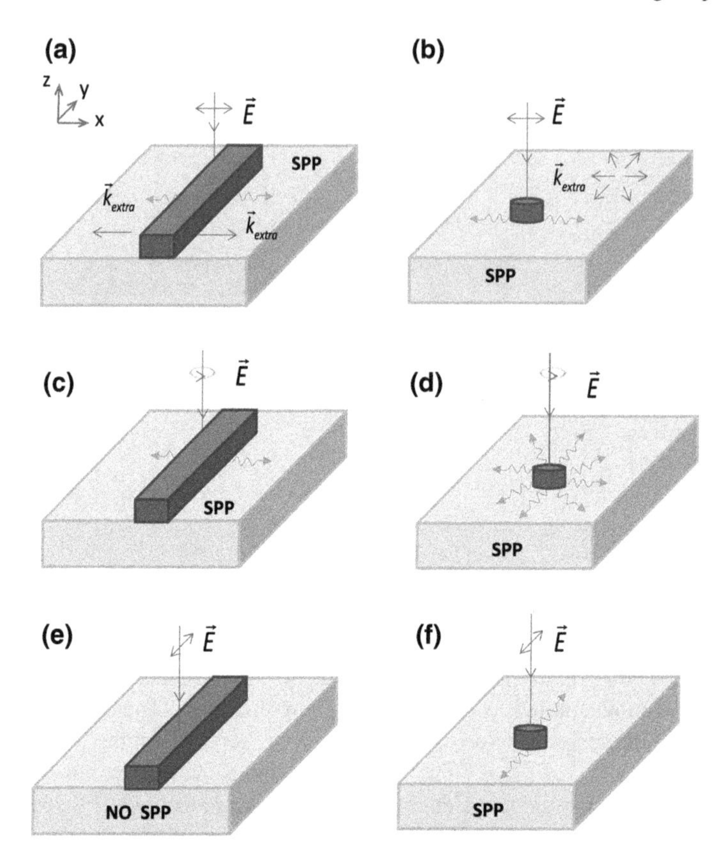

Fig. 2.7 SPP coupling using a one-dimensional defect, a rectangular protrusion (**a**), (**c**) and (**e**), and a circular (2-D) one (**b**), (**d**) and (**f**). In the upper panels the polarization of incident light is in the x direction (**a**) and (**b**), while in the middle ones the light is circularly polarized (**c**) and (**d**) and in the lower ones it is linearly polarized along the y direction (**e**) and (**f**). For the 1-D defect (*left panels*), the extra momentum is along the x direction, while for the circular ones (*right*) the extra momentum is provided in all the directions of the plane

For more complex geometries such as thin metallic films, or multilayers, the dispersion relation of the SPP at the interfaces can be obtained in a similar way to the one described here. However, usually there is not an analytical expression but it has to be solved numerically. An example of two of those geometries is a thin metal layer between two identical dielectrics (known as insulator-metal-insulator, IMI, configuration), or a thin insulating (dielectric) layer between two metals (MIM configuration) [4, 7, 37, 38]. Those symmetric configurations lead to interesting properties when reducing the size of the middle layer, since for a sufficiently small middle layer the SPP is split in two SPP modes, each one with its own characteristics. For the IMI situation, the confinement of one of the SPP modes decreases a lot when decreasing the metal thickness but its propagation increases, and it is called the long-range SPP [4, 7]. For the MIM situation, on the other hand, the possibility to guide the strongly confined plasmon along the insulating layer is presented, and it serves as a PGW [4, 37, 38]. However, although these configurations have different applications and interesting physical effects, they are out of the scope of this thesis.

2.2 Active Plasmonics

As mentioned before, plasmonic circuitry is a flourishing topic of research [28, 30]. This is due to its compatibility with traditional Si-CMOS technology, adding the scaling beyond the diffraction limit of light that SPPs provide. Plasmonic circuits will lead to fast and small devices, and reduced optical power requirements. Nevertheless, in order to get those plasmonic circuits, we have to solve two main fundamental difficulties that SPPs pose: losses and external manipulation. Regarding losses, there are several works dealing with them by means of gaining media [31, 32]. In this thesis we will focus on the external manipulation of SPPs, which is denoted as active plasmonics [39].

Achieving active plasmonic configurations is a critical step to really endow plasmonic systems with full capacity to develop nanophotonic chips, as it will allow the realization of fundamental components such as modulators, switches or active multiplexors, couplers and add-drop filters. This requires systems where the plasmon properties can be more or less rapidly modulated by an external parameter. So far, the ability of parameters such as temperature [39–43], voltage [44–48], or optical signals [49–54] to act on SPPs and develop some sort of active device has already been demonstrated. In all cases, the mechanism underlying the modification of the system response is either the control of absorption [40, 44, 47, 49, 50] or the modification of the material refractive index and thus of the SPP wavevector [42, 45, 55, 56] (or even a combination of both in some specific cases [51, 52]).

Each proposal has its marketable advantages, as well as its weaknesses and, depending on the application, it will be better to choose one external agent or another. There are several aspects on which one should focus:

- switching speed: the speed at which the SPP can be modified, which implies not only the rate of the applied control signal but also the dynamic response of the material;
- magnitude of the modulation: it is fundamental to quantify the effect that we produce on the SPP and on the final signal to design an appropriate device;
- feasibility: this implies both the size and the simplicity of the device relevant to make it competitive;
- cost: the prize of the device is always an aspect to be considered.

In general, active devices with temperature control are the slowest, with a switching speed that could be of milliseconds [41]. However, large modulations can be achieved in this case, and their implementation has been already demonstrated [41, 42, 57], with reasonable electrical power requirements. Optical or voltage control depend on each particular case; however, they are not the perfect solution either.

Usually, when they are excellent at modulating or at being integrated in a circuit, they are slow or viceversa. For example, in Ref. [46] they use a dielectric loaded SPP waveguide doped with molecules that act as an electro-optic material. They achieve modulations of about 16 % in a reduced metallic resonator covered by a dielectric by applying an electric field, but the time response is of the order of seconds. In Ref. [47], on the other hand, they use photocromic molecules that show a reversible switch from transparent to absorbing when they are illuminated in a polymethyl methacrylate (PMMA) matrix. They get considerable modulation (about 70 %), but the particles and their response degrade after several switchings.

The magnetic field is another interesting candidate to control SPPs, since it is known to affect the optical properties of materials (magnetooptics). Indeed, a magnetic field applied in the appropriate direction can modify the SPP dispersion relation [58–61]. This thesis deals with the use of a magnetic field within the context of active plasmonics and the potential development of integrated modulators based on this [55, 56, 62], as well as the possibility to improve the performance of surface plasmon sensing devices by adding magnetic modulation [63]. We have chosen the magnetic field as external agent mainly because it is fast and easy to apply. The intrinsic switching speed of magnetization in ferromagnetic materials can achieve values as high as femtoseconds [64], that would be one of the fastest for active plasmonics. Besides, with the appropriate materials, magnetic fields of about 20 mT are enough to induce a measurable magnetic response on the system, which is a reasonable value to achieve both in a laboratory and in a device. Finally, the dimensions of the modulator can be of about tens of microns, an appropriate size for circuit implementation.

2.3 Magnetoplasmonics

Magnetoplasmonics deals with the intertwined properties of magnetooptical activity and SP [65], that is, the interaction between surface plasmon resonances and magnetooptics. There are works from the 70s and the 80s already describing this interaction for semiconductors and metals [58, 66, 67]. There are two main aspects to be considered here. First, the SP resonances are able to enhance the MO Kerr or Faraday effect. This enhancement of the MO effects is due to the enhancement of the electromagnetic field associated with SP excitation. At the same time, since the MO effects are usually normalized changes in the reflectivity (or transmission), the SP related drop in the reflectivity can also enhance this MO activity [65]. LSPs are widely efficient due to their strong nanoantenna effects, leading to an important increase on the MO activity [68–70]. This enhancement of the MO properties has been used for example to obtain, by using gold nanodisks, the magnetooptical constants of Au [71], whose magnetic effects are usually so small that can not be measured unless using exceptional equipment. By contrast, SPPs main effects on the MO response are related to the reduction of the reflectivity [72, 73]. Nevertheless, in this thesis we are interested in the other aspect of the SP resonance / MO activity interaction: under certain conditions, we can use the magnetic field to modify the plasmonic behavior.

When dealing with magnetic fields and optical properties, we need to describe the optical properties by means of a dielectric tensor. In absence of magnetic field, this tensor is (or can be converted into) a diagonal one. When a magnetic field is applied the light electric fields are rotated and this dielectric tensor becomes-non diagonal. Those non-diagonal elements, usually denoted magnetooptic constants, $\varepsilon_{ij} \neq 0$, are larger or smaller depending on the material used. The general dielectric tensor is [74]:

$$\hat{\varepsilon} = \begin{pmatrix} \varepsilon_{xx} & \varepsilon_{xy} & \varepsilon_{xz} \\ -\varepsilon_{xy} & \varepsilon_{yy} & -\varepsilon_{yz} \\ -\varepsilon_{xz} & \varepsilon_{yz} & \varepsilon_{zz} \end{pmatrix}, \tag{2.13}$$

being ε_{ii} the dielectric (optical) constants of the material at the i direction. The magnetooptical constants can also be expressed by means of the magnetooptical Q factor, a parameter that relates the magnetooptical constants with the optical ones: $Q = i\frac{\varepsilon_{ij}}{\varepsilon_{ii}}$. magnetooptic constants depend on the magnetic nature of the used material. For diamagnetic and paramagnetic materials, those constants are proportional to the magnetic field applied and quite small [65]. Therefore to achieve large MO constants in those materials, a large magnetic field is necessary. For ferromagnetic materials, on the other hand, these MO constants are much bigger and proportional to the magnetization, which means that we can saturate magnetically these materials and that larger MO constants are achieved for lower magnetic fields [65].

For magnetoplasmonics, we want to combine SPPs and magnetic effects. Then we have to analyze carefully both the optical and MO properties of the involved materials. The quality of the SPPs response is determined by the optical losses, being noble metals the ones with narrower resonance curves and therefore larger propagation distances. However, noble metals are diamagnetic, and thus have very small MO constants at reasonable field values, although their SPP properties are the best ones. Ferromagnetic materials (Co, Fe, Ni), on the other hand, have large ε_{ij} values at small magnetic fields (proportional to their magnetization), but they are optically too absorbent, reducing notably the SPP propagation distance and limiting its use in a photonic device.Therefore we need a smart combination of both types of materials in order to take advantage of the good properties of both, and minimize the drawbacks (this will be discussed in more detail in the next section).

In magnetooptics (and in magnetoplasmonics in particular), it has to be mentioned that the relative orientation of the magnetic field and the optical incidence plane is essential, since this determines the particular non-zero ε_{ij} components [58, 74]. There are three different configurations, shown in Fig. 2.8, denoted as transverse, polar, and longitudinal [74] in the well known MO effects in reflection (Kerr) and transmission (Faraday) [75, 76].[3] The non-zero MO constants that appear for each configuration are those not involving the direction of the magnetic field, and their effect on a wave is that they couple the corresponding electromagnetic fields involved. We will

[3]Some details of the Kerr effect for two configurations can be found in the Appendix A.2.

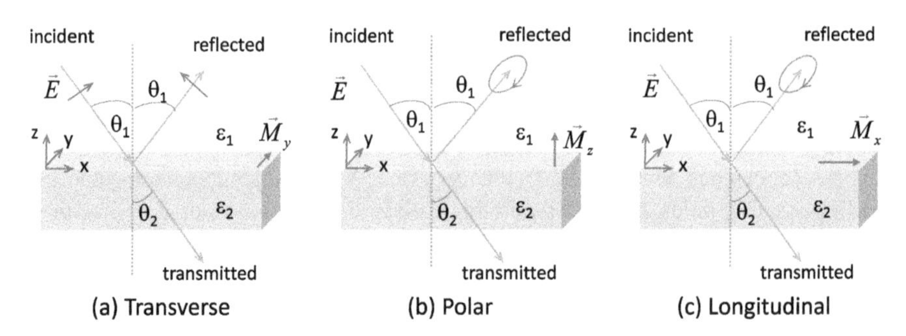

Fig. 2.8 Different configurations of the Faraday (transmission) and Kerr (reflection) effects. ε_1 represents the "non-magnetic" material and ε_2 the "magnetic" one. **a** Transverse configuration: the magnetic field is applied in the sample plane and perpendicular to the plane of incidence. **b** Polar configuration: the magnetic field is perpendicular to the sample plane. **c** Longitudinal configuration: the magnetic field is parallel to both the sample plane and the incidence one

consider a SPP or a wave in the XZ plane and the coordinates system of Fig. 2.8. In the case of the polar configuration, magnetization along the z direction, $\varepsilon_{xy} \neq 0$. This implies a conversion between the E_x and the E_y component of the field, and as a consequence a conversion of polarization. However, this is not an appropriate configuration to work with SPPs, as the plasmon do not have any E_y component (it is a TM mode). Therefore, it will radiate light causing losses, which is not the desired effect. This "depolarization" of the plasmon happens also in the longitudinal configuration ($\varepsilon_{yz} \neq 0$). The suitable MO configuration when dealing with SPPs in then the transverse one ($\varepsilon_{xz} \neq 0$). In this case, there will be a conversion between E_x and E_z, i.e. the wave stays a TM wave. Since for SPPs these field components are related by Eqs. 2.3–2.5, we can intuitively see that their relative variation gives rise to a modification of the SPP wavevector, as it will be explained in detail in the following section.

2.4 Magnetic Modulation of the SPP Dispersion Relation

The same procedure detailed in Sect. 2.1.1 to find the SPP dispersion relation in an isotropic metal/dielectric interface can be applied in the case of an interface where one of the media (or both) is magnetooptic, just by introducing the appropriate dielectric tensor to describe that medium. For the transverse configuration, and keeping the same coordinate system that for Figs. 2.1 and 2.8, this finally results in a magnetic field modification of the SPP wavevector k_x, introducing a term in the SPP dispersion relation [58, 65], linear with the normalized magnetization m:

$$k_x(\vec{B}) = k_x^0 + \Delta k_x \cdot m, \tag{2.14}$$

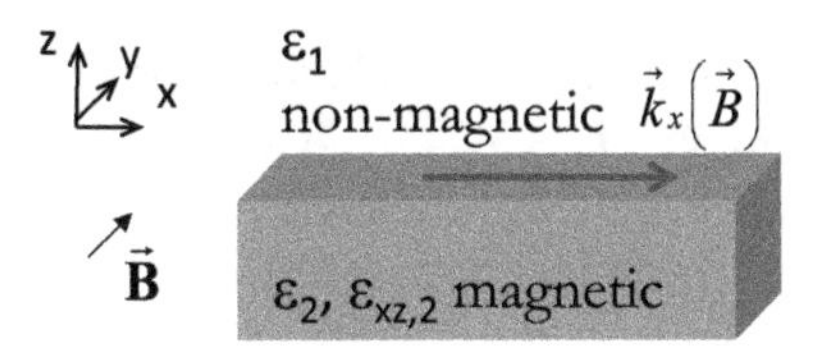

Fig. 2.9 Schema of the single interface configuration for which the magnetic modulation of the SPP wavevector is obtained. ε_1 represents the non-magnetic material whereas ε_2 is the magnetic one

where k_x^0 is the SPP wavevector when there is no magnetic field applied, and $m = \frac{M_y}{M_{sat}}$, therefore $-1 \le m \le 1$. The maximum effect can be achieved applying magnetic fields large enough to saturate the sample so that m is either 1 or -1. The proportionality factor of the magnetic modulation of the SPP wavevector, Δk_x, is related to the dielectric constants of both materials at the interface, including the magnetooptical ones. For the simplest case, a single interface with two semi-infinite materials, a dielectric and a metal, the expression for the modulation can be analytically obtained as a function of the dielectric constants of the interface materials (ε_1, ε_2) applying the scattering matrix formalism [77]. In this case, considering that the magnetic material is the medium 2 (see Fig. 2.9), and assuming that $\varepsilon_{xz,2} \ll \varepsilon_2$, the magnetic modulation can be written as [60, 78]:

$$\Delta k_x = -k_x^0 \frac{\varepsilon_{xz,2}}{\sqrt{\varepsilon_1 \varepsilon_2}(1 - \frac{\varepsilon_2^2}{\varepsilon_1^2})} + O(\varepsilon_{zx}^2). \tag{2.15}$$

As the metal dielectric constant is complex, this modification occurs in both the real and imaginary part of k_x: k_x^r (Δk_x^r) and k_x^i (Δk_x^i), respectively. k_x^r and k_x^i represent different physical properties. Then, the magnetic field offers the opportunity of designing active plasmonic systems based on the control of absorption if Δk_x^i dominates, or in effective refractive index modification if Δk_x^r does. Moreover, as can be seen in Eq. 2.14, the SPP modification under an applied magnetic field is non-reciprocal ($k_x(B_0) \neq -k_x(B_0)$) [65]. This makes this kind of systems very interesting for the design of integrated optical isolators [59, 79–82], which is another fundamental piece for the development of nanophotonic circuits.

2.4.1 MO Activity: In the Metal or in the Dielectric?

Equation 2.15 shows that there is a substantial difference in the value of the magnetic field modulation of the SPP when the magnetic material is the metal or the dielectric, since $|\varepsilon_m| > \varepsilon_d$. If we write the Eq. 2.15 as a function of the MO factor Q:

$$\Delta k_x = k_x^0 \frac{i Q_2}{1 - R_{21}^2} \sqrt{R_{21}}, \tag{2.16}$$

where $R_{21} = \frac{\varepsilon_2}{\varepsilon_1}$. Let's assume that we could have a metal and a dielectric with the same magnetooptical factor Q_2 ($Q = i\frac{\varepsilon_{xz}}{\varepsilon_{xx}}$), and representative values for the optical constants of both materials at a given wavelength to allow plasmons: $\varepsilon_m = -20$ and $\varepsilon_d = 2$. When the magnetic material is the metal, *metal* = 2 and *dielectric* = 1, then $\varepsilon_m = \varepsilon_2 = -20$, $\varepsilon_d = \varepsilon_1 = 2$, and $R_{21} = -10$:

$$\Delta k_x = k_x^0 \frac{iQ}{1 - 10^2}\sqrt{-10} \approx k_x^0 \frac{Q}{30} \tag{2.17}$$

On the other hand, if the magnetic material is the dielectric, *metal* = 1 and *dielectric* = 2, $\varepsilon_m = \varepsilon_1 = -20$, $\varepsilon_d = \varepsilon_2 = 2$, and $R_{21} = -0.1$:

$$\Delta k_x = k_x^0 \frac{iQ}{1 - \frac{1}{10}^2}\sqrt{\frac{1}{-10}} \approx k_x^0 \frac{Q}{3} \tag{2.18}$$

As it can be seen, for the same Q factor the relative modulation of the SPP wavevector is almost 10 times larger if the magnetic material is the dielectric than if it is the metal. This is mainly due to the difference in the mentioned spreading or confinement of the electromagnetic field inside the metal or the dielectric (Sect. 2.1.1). When a magnetic field is applied, the modulation will be larger if the ferromagnetic material is the one on which there is more electromagnetic field spread. In this example, the distribution of the electromagnetic field of the plasmon along the two materials (such as in Fig. 2.4) is going to change only in position depending on which material is the metal, but not in relative values. As there is more SPP electromagnetic field spread along the dielectric than along the metal, if the ferromagnetic material is the dielectric the SPP would feel more the magnetic influence and the modulation will be larger.

However, it also happens that, for actual materials, the Q factor is larger for a ferromagnetic metal (Co, Fe or Ni) than for a ferromagnetic dielectric (garnets or ferrites). For example, at 2 eV the real part of the Q factor for Co [83, 84], an iron garnet [85],[4] and a Co ferrite[5] are −0.019, −0.003 and −0.001, respectively. So the example shown before is not a realistic result, and results using actual dielectric or metal ferromagnetic materials are similar. However, this helps us to illustrate the main aspects that concern the appropriate choice of materials for the modulation of the SPP wavevector. Some works have demonstrated SPP modulation by using Au/ferromagnetic dielectric interfaces [78, 86], but in this thesis we have chosen the ferromagnetic metal because it can be easily fabricated and the magnetic fields needed to achieve saturation are smaller. Nevertheless, the metallic part cannot be just a ferromagnetic material since, as it has been said, it lacks of competitive plasmonic properties. An appropriate trade-off is then working with multilayers combining noble and ferromagnetic metals [55].

[4] $Y_{3-x}Bi_xFe_5O_{12}$, with $x = 1.07$.

[5] Experimentally obtained at IMM.

2.4.2 Thin Ferromagnetic Layer

However, this is an ideal situation, and it has been explained the convenience of using metallic multilayers combining ferromagnetic and noble metals. The simplest multilayer relevant here consists of a system with a single metal/dielectric interface and a *thin ferromagnetic metallic layer* inserted in the noble metallic material (see Fig. 2.10). We have obtained an expression for the modulation of k_x applying Maxwell equations in a system consisting of a trilayer and a semi-infinite dielectric, taking the dielectric tensor to account for the optical and magneto-optical response of the ferromagnetic layer and in the approximation that this ferromagnetic layer is very thin [55, 87]. Then the magnetic field induced variation of the SPP wavevector becomes [55, 56]:

$$\Delta k_x = \frac{2t_{ferr}(k_0\varepsilon_d\varepsilon_m)^2 i\varepsilon_{xz}}{(\varepsilon_d+\varepsilon_m)(\varepsilon_d^2-\varepsilon_m^2)\varepsilon_{ferr}} e^{-2h\cdot k^i_{z,m}}, \tag{2.19}$$

or, by considering $|\varepsilon_m| \gg |\varepsilon_d|$:

$$\Delta k_x \approx \frac{i2t_{ferr}k_0^2\varepsilon_d^2\varepsilon_{xz}}{-\varepsilon_m\varepsilon_{ferr}} e^{-2h\cdot k^i_{z,m}} = \frac{i2t_{ferr}k_0^2\varepsilon_d^2\varepsilon_{xz}}{-\varepsilon_m\varepsilon_{ferr}} e^{-h/\delta_m}, \tag{2.20}$$

where $k^i_{z,m}$ is the imaginary part of the vertical component of the SPP wavevector in the noble metal layer, δ_m is the penetration distance of the SPP electromagnetic field in the metal, h is the thickness of the upper noble metal layer, t_{ferr} is the thickness of the ferromagnetic layer, k_0 is the wavevector of the incident light in vacuum, ε_m is the (isotropic) complex dielectric permittivity of the noble metal, ε_d is the dielectric constant of the dielectric medium (supposed real), and ε_{ferr} and ε_{xz} are the diagonal and non-diagonal elements of the complex dielectric tensor for the ferromagnetic layer. These parameters are shown in Fig. 2.10. The feasibility of using this noble metal layer with a thin ferromagnetic metal layer inserted for SPP modulation has been demonstrated in Ref. [55]. The system analyzed there consist of a 200 nm gold layer with a thin layer of cobalt inside, forming a Au/Co/Au trilayer. Gold provides the suitable SPP properties, it presents no (or negligible) degradation with time and its functionalization processes are well known, so that it can also be implemented in a biological application. Co is the ferromagnetic metal chosen due to its large enough

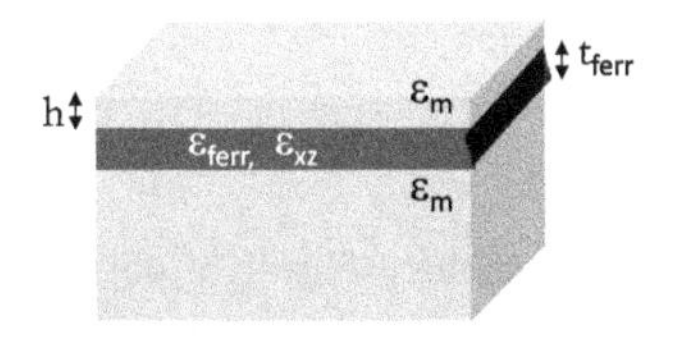

Fig. 2.10 Schema of a single metal/dielectric interface configuration with a noble metal/thin ferromagnetic metal/noble metal multilayer

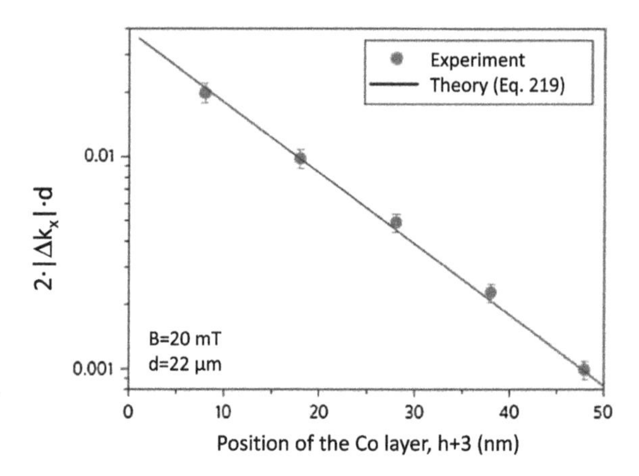

Fig. 2.11 Extracted from Temnov et al. [55] Evolution of the magnetic modulation of the SPP wavevector as a function of the position of the Co layer for a Au/Co/Au trilayer. It shows an exponential decay with a penetration depth of 13 nm

MO constants and its relatively low saturation magnetic field. The Co layer thickness is of 6 nm, so the "thin layer approximation" can be applied [87]. The position of the Co layer is varied from 5 to 45 nm. In this work, Temnov et al., by means of plasmonic interferometry (which will be explained in detail in the next chapter) have calculated the modulation of the SPP wavevector Δk_x under an applied magnetic field. They have obtained that the magnetic modulation of the SPP decays exponentially with the Co layer position, in agreement with Eqs. 2.19 and 2.20. This result illustrates the relation between the spreading of the electromagnetic field inside the MO active layer and the modulation of the SPP wavevector. Figure 2.11 shows the evolution of the magnetic field modulated SPP wavevector as a function of the Co depth. As it can be seen, it presents an exponential decay and, in fact, the skin depth of the SPP into the trilayer is experimentally obtained using Eq. 2.20, being of about 13 nm. Moreover, these MP interferometers do not only act as probes for the SPP wavevector modulation but they also show promising properties as an optical switch, so that they can act as a device *per se*. Indeed, modulation depths up to 2 % are demonstrated in that work, which seems promising for potential applications. These results have then encouraged us to undertake a deeper study of this kind of MP interferometers with noble metal/thin ferromagnetic metal/noble metal multilayers, in order to understand all the parameters that influence the values of the SPP wavevector modulation as well as the possibilities to enhance it. The basis of magnetoplasmonic interferometry, both as a device as well as a method to extract Δk_x, are explained in the next chapter.

References

1. www.britishmuseum.org/explore/highlights/highlight_objects/pe_mla/t/the_lycurgus_cup.aspx
2. J.M.C. Garnett, Philos. Trans. R. Soc. Lond. **203**, 385 (1904)
3. G. Mie, Ann. Phys. **25**, 377 (1908)
4. S.A. Maier, *Plasmonics: Fundamentals and Applications* (Springer, Berlin, 2007)

5. C.F. Bohren, D.R. Huffman, *Absorption and Scattering of Small Particles* (Wiley-VCH, New York, 2004)
6. A.V. Zayats, I.I. Smolyaninov, J. Opt. A Pure Appl. Opt. **5**, 16–50 (2003)
7. H. Raether, *Surface Plasmons* (Springer, Berlin, 1986)
8. A. Sommerfeld, Annalen der Physic **67**, 233–290 (1899)
9. J. Zenneck, Annalen der Physic **23**, 846–866 (1907)
10. R.W. Wood, Philos. Mag. **4**, 396 (1902)
11. U. Fano, J. Opt. Soc. Am. **31**, 213–222 (1941)
12. D. Pines, Rev. Mod. Phys. **28**, 184–198 (1956)
13. R.H. Ritchie, Phys. Rev. **106**, 874–881 (1957)
14. A. Otto, Zeitschrift for Physik **216**, 398–410 (1968)
15. E. Kretschmann, H.Z. Raether, Natruf **23A**, 2135 (1968)
16. G. Baffou, R. Quidant, Laser Photonics Rev. **7**, 171–187 (2013)
17. M. Ozaki, J.-I. Kato, S. Kawata, Science **332**, 218–220 (2011)
18. Z. Jacob, MRS Bull. **37**, 761–767 (2012)
19. A. Otto, I. Mrozek, H. Grabhorn, W. Akemann, J. Phys. Condens. Matter **4**, 1143 (1992)
20. F.J. Garcia-Vidal, J.B. Pendry, Phys. Rev. Lett. **77**, 1163–1166 (1996)
21. K. Kneipp, Y. Wang, H. Kneipp, L.T. Perelman, I. Itzkan, R.R. Dasari, M.S. Feld, Phys. Rev. Lett. **78**, 1667–1670 (1997)
22. T.W. Ebbesen, H.J. Lezec, H.F. Ghaemi, T. Thio, P.A. Wolff, Nature **391**, 667–669 (1998)
23. L. Martin-Moreno, F.J. Garcia-Vidal, H.J. Lezec, K.M. Pellerin, T. Thio, J.B. Pendry, T.W. Ebbesen, Phys. Rev. Lett. **86**, 1114–1117 (2001)
24. J. Homola, Chem. Rev. **108**, 462–493 (2008)
25. J.N. Anker, W.P. Hall, O. Lyandres, N.C. Shah, J. Zhao, R.P. Van Duyne, Nat. Mater. **7**, 442 (2008)
26. W.L. Barnes, A. Dereux, T.W. Ebbesen, Nature **424**, 824–830 (2003)
27. E. Ozbay, Science **311**, 189–193 (2006)
28. T.W. Ebbesen, C. Genet, S.I. Bozhevolnyi, Phys. Today **61**, 44–50 (2008)
29. T. Holmgaard, Z. Chen, S.I. Bozhevolnyi, L. Markey, A. Dereux, A.V. Krasavin, A.V. Zayats, Opt. Express **16**, 13585–13592 (2008)
30. D.K. Gramotnev, S.I. Bozhevolnyi, Nat. Photonics **4**, 83–94 (2010)
31. J. Grandidier, G. Colas des Francs, S. Massenot, A. Bouhelier, L. Markey, J.-C. Weeber, C. Finot, A. Dereux, Nano Lett. **9**, 2935–2939 (2009)
32. P. Berini, I. De Leon, Nat. Photonics **6**, 16–24 (2012)
33. P.B. Johnson, R.W. Christy, Phys. Rev. B **6**, 4370–4379 (1972)
34. H. Ditlbacher, J.R. Krenn, N. Felidj, B. Lamprecht, G. Schider, M. Salerno, A. Leitner, F.R. Aussenegg, Appl. Phys. Lett. **80**, 404–406 (2002)
35. H. Ditlbacher, J.R. Krenn, A. Hohenau, A. Leitner, F.R. Aussenegg, Appl. Phys. Lett. **83**, 3665–3667 (2003)
36. J. Renger, S. Grafström, L.M. Eng, Phys. Rev. B **76**, 045431 (2007)
37. J.A. Dionne, L.A. Sweatlock, H.A. Atwater, A. Polman, Phys. Rev. B **73**, 035407 (2006)
38. P. Ginzburg, D. Arbel, M. Orenstein, Opt. Lett. **31**, 3288–3290 (2006)
39. A.V. Krasavin, N.I. Zheludev, Appl. Phys. Lett. **84**, 1416–1418 (2004)
40. A.V. Krasavin, K.F. MacDonald, N.I. Zheludev, A.V. Zayats, Appl. Phys. Lett. **85**, 3369–3371 (2004)
41. T. Nikolajsen, K. Leosson, S.I. Bozhevolnyi, Appl. Phys. Lett. **85**, 5833–5835 (2004)
42. J. Gosciniak, S.I. Bozhevolnyi, T.B. Andersen, V.S. Volkov, J. Kjelstrup-Hansen, L. Markey, A. Dereux, Opt. Express **18**, 1207–1216 (2010)
43. J. Gosciniak, S.I. Bozhevolnyi, Sci. Rep. **3**, 1803 (2013)
44. J.A. Dionne, K. Diest, L.A. Sweatlock, H.A. Atwater, Nano Lett. **9**, 897–902 (2009)
45. M.J. Dicken, L.A. Sweatlock, D. Pacifici, H.J. Lezec, K. Bhattacharya, H.A. Atwater, Nano Lett. **8**, 4048–4052 (2008)
46. S. Randhawa, S. Lacheze, J. Renger, A. Bouhelier, R.E. de Lamaestre, A. Dereux, R. Quidant, Opt. Express **20**, 2354–2362 (2012)

47. A. Agrawal, C. Susut, G. Stafford, U. Bertocci, B. McMorran, H.J. Lezec, A.A. Talin, Nano Lett. **11**, 2774–2778 (2011)
48. D.C. Zografopoulos, R. Beccherelli, Plasmonics **8**, 599–604 (2013)
49. D. Pacifici, H.J. Lezec, H.A. Atwater, Nat. Photonics **1**, 402–406 (2007)
50. R.A. Pala, K.T. Shimizu, N.A. Melosh, M.L. Brongersma, Nano Lett. **8**, 1506–1510 (2008)
51. K.F. MacDonald, Z.L. Samson, M.I. Stockman, N.I. Zheludev, Nat. Photonics **3**, 55–58 (2009)
52. V.V. Temnov, K. Nelson, G. Armelles, A. Cebollada, T. Thomay, A. Leitenstorfer, R. Bratschitsch, Opt. Express **17**, 8423–8432 (2009)
53. M.P. Nielsen, A. Elezzabi, Opt. Express **21**, 20274–20279 (2013)
54. J. Chen, Z. Li, J. Xiao, Q. Gong, Plasmonics **8**, 233–237 (2013)
55. V.V. Temnov, G. Armelles, U. Woggon, D. Guzatov, A. Cebollada, A. Garcia-Martin, J.M. Garcia-Martin, T. Thomay, A. Leitenstorfer, R. Bratschitsch, Nat. Photonics **4**, 107–111 (2010)
56. D. Martin-Becerra, J.B. Gonzalez-Diaz, V. Temnov, A. Cebollada, G. Armelles, T. Thomay, A. Leitenstorfer, R. Bratschitsch, A. Garcia-Martin, M.U. Gonzalez, Appl. Phys. Lett. **97**, 183114 (2010)
57. A.L. Lereu, A. Passian, J.-P. Goudonnet, T. Thundat, T.L. Ferrell, Appl. Phys. Lett. **86**, 154101 (2005)
58. R.F. Wallis, J.J. Brion, E. Burstein, A. Hartstein, Phys. Rev. B **9**, 3424–3437 (1974)
59. Z. Yu, G. Veronis, Z. Wang, S. Fan, Phys. Rev. Lett. **100**, 023902 (2008)
60. V.I. Belotelov, D.A. Bykov, L.L. Doskolovich, A.N. Kalish, A.K. Zvezdin, J. Opt. Soc. Am. B **26**, 1594–1598 (2009)
61. J.B. Gonzalez-Diaz, A. Garcia-Martin, G. Armelles, J.M. Garcia-Martin, C. Clavero, A. Cebollada, R.A. Lukaszew, J.R. Skuza, D.P. Kumah, R. Clarke, Phys. Rev. B **76**, 153402 (2007)
62. D. Martin-Becerra, V.V. Temnov, T. Thomay, A. Leitenstorfer, R. Bratschitsch, G. Armelles, A. Garcia-Martin, M.U. Gonzalez, Phys. Rev. B **86**, 035118 (2012)
63. D. Martin-Becerra, G. Armelles, M.U. Gonzalez, A. Garcia-Martin, New J. Phys. **15**, 085021 (2013)
64. A.V. Kimel, A. Kirilyuk, P.A. Usachev, R.V. Pisarev, A.M. Balbashov, T. Rasing, Nature **435**, 655 (2005)
65. G. Armelles, A. Cebollada, A. Garcia-Martin, M.U. Gonzalez, Adv. Opt. Mater. **1**, 10–35 (2013)
66. K.W. Chiu, J.J. Quinn, Phys. Rev. Lett. **29**, 600–603 (1972)
67. W. Reim, O.E. Husser, J. Schoenes, E. Kaldis, P. Wachter, K. Seiler, J. Appl. Phys. **55**, 2155–2157 (1984)
68. P.M. Hui, D. Stroud, Appl. Phys. Lett. **50**, 950–952 (1987)
69. J.B. Gonzalez-Diaz, A. Garcia-Martin, J.M. Garcia-Martin, A. Cebollada, G. Armelles, B. Sepulveda, Y. Alaverdyan, M. Kll, Small **4**, 202–205 (2008)
70. L. Wang, C. Clavero, Z. Huba, K. Carroll, E.E. Carpenter, D. Gu, R.A. Lukaszew, Nano Lett. **11**, 1237–1240 (2011)
71. B. Sepulveda, J.B. Gonzalez-Diaz, A. Garcia-Martin, L.M. Lechuga, G. Armelles, Phys. Rev. Lett. **104**, 147401 (2010)
72. E. Ferreiro-Vila, J.B. Gonzalez-Diaz, R. Fermento, M.U. Gonzalez, A. Garcia-Martin, J.M. Garcia-Martin, A. Cebollada, G. Armelles, D. Meneses-Rodriguez, E. Muoz Sandoval, Phys. Rev. B **80**, 125132 (2009)
73. G. Ctistis, E. Papaioannou, P. Patoka, J. Gutek, P. Fumagalli, M. Giersig, Nano Lett. **9**, 1–6 (2009)
74. A.K. Zvezdin, V.A. Kotov, *Modern Magnetooptics and Magnetooptical Materials* (IOP Publishing Ltd, Bristol, 1997)
75. J. Kerr, Philos. Mag. **3**, 332 (1877)
76. M. Faraday, Philos. Trans. R. Soc. Lond. **136**, 1–20 (1846)
77. B. Caballero, A. Garcia-Martin, J.C. Cuevas, Phys. Rev. B **85**, 245103 (2012)
78. V.I. Belotelov, I.A. Akimov, M. Pohl, V.A. Kotov, S. Kasture, A.S. Vengurlekar, A.V. Gopal, D.R. Yakovlev, A.K. Zvezdin, M. Bayer, Nat. Nanotechnol. **6**, 370–376 (2011)
79. J.B. Khurgin, Appl. Phys. Lett. **89**, 251115 (2006)

80. W. Van Parys, B. Moeyersoon, D. Van Thourhout, R. Baets, M. Vanwolleghem, B. Dagens, J. Decobert, O. Le Gouezigou, D. Make, R. Vanheertum, L. Lagae, Appl. Phys. Lett. **88**, 071115 (2006)
81. B. Sepulveda, L. Lechuga, G. Armelles, J. Lightwave Technol. **24**, 945–955 (2006)
82. J. Montoya, K. Parameswaran, J. Hensley, M. Allen, R. Ram, J. Appl. Phys. **106**, 023108 (2009)
83. P.B. Johnson, R.W. Christy, Phys. Rev. B **9**, 5056–5070 (1974)
84. G.S. Krinchik, J. Appl. Phys. **35**, 1089–1092 (1964)
85. V. Doormann, J.-P. Krumme, H. Lenz, J. Appl. Phys. **68**, 3544–3553 (1990)
86. M. Pohl, L.E. Kreilkamp, V.I. Belotelov, I.A. Akimov, A.N. Kalish, N.E. Khokhlov, V.J. Yallapragada, A.V. Gopal, M. Nur-E-Alam, M. Vasiliev, D.R. Yakovlev, K. Alameh, A.K. Zvezdin, M. Bayer, New J. Phys. **15**, 075024 (2013)
87. J.F. Torrado, J.B. Gonzalez-Diaz, A. Garcia-Martin, G. Armelles, New J. Phys. **15**, 075025 (2013)

Chapter 3
Magnetoplasmonic Interferometry

This chapter contains all the information about plasmonic and magnetoplasmonic interferometry. It is explained how it works, the basic parameters implied and the methodology used to obtain information about the SPP magnetic modulation, as well as its capability as a device.

Interferometry is known as a very sensitive and reliable technique both at the micro- and macro-scale, and it has been applied during many years and in many areas such as astrophysics [1, 2], imaging [3–5], sensing [6], or integrated photonics [7, 8]. One of the most famous experiments regarding interferometry took place in 1887, when Michelson and Morley built an interferometer to detect the presence of ether in the space and see its effect on the speed of light [9]. The experiment did not succeed, resulting in that there was no ether at all, but the use of this kind of interferometer to have very accurate results was then well established. Recently, the concept of wave interference has been extended to surface plasmons, and it has given rise to the so-called plasmonic interferometers [10–15]. In fact, plasmonic interferometry has been demonstrated, both theoretically and experimentally, for sensing [16–22] as well as for integrated devices [15, 23, 24].

The ability of interferometry to detect and be sensitive to small modifications of the light and of the SPP properties, makes it a straightforward tool to combine it with an external agent and detect active control of SPPs [12, 25, 26]. Therefore, a SPP interferometer coupled with a magnetic field will constitute our magnetically modulated SPP interferometer [27]. Moreover, this MP interferometer can be used not only to measure the magnetic field induced modification of the SPP wavevector, but also as a device itself, a modulator or an optical switch, as it will be shown.

In this chapter, we are going to describe the experimental and theoretical details of magnetoplasmonic interferometry. We will first analyze MP interferometers as a technique for measuring the magnetic modulation of the SPP wavevector Δk_x, and after that we will evaluate their characteristics as a device (modulator) for a circuit.

D. Martín Becerra, *Active Plasmonic Devices*, Springer Theses,
DOI 10.1007/978-3-319-48411-2_3

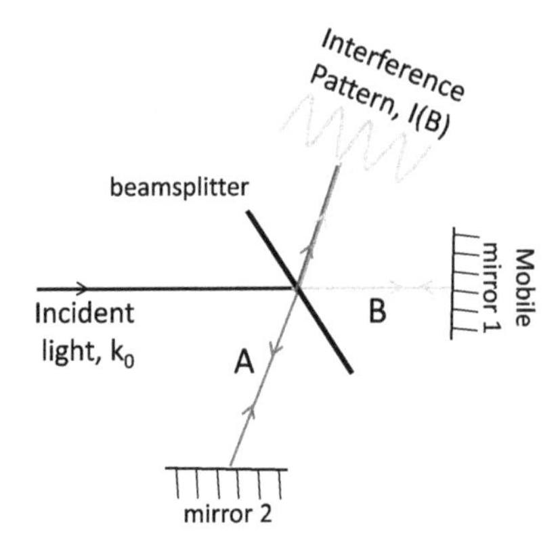

Fig. 3.1 Schema of a basic Michelson interferometer. The incident light, of wavevector k_0, is split into two beams that travel different distances A and B. When the two beams are superimposed, each of them has covered a different optical path, resulting in a interference. Since one of the mirrors can be moved, the distance B will vary, therefore leading to a complete interference pattern

3.1 Interferometry

There are several types of interferometers, but all of them consist in the superposition of two waves and the analysis of the pattern that is formed due to the different paths traveled by the two waves [28]. Usually one of the waves is fixed as a reference, while the other one travels different paths, depending on the properties that are studied. For a basic Michelson interferometer, outlined in Fig. 3.1, the intensity of the interference can be written as:

$$I_T = |E_A + E_B|^2 = I_A + I_B + 2\sqrt{I_A I_B}\cos[k_0(A - B)], \tag{3.1}$$

where A and B refer to the two arms of the interferometer, and k_0 is the light wavevector in air. Depending on the difference between the two optical paths, $k_0(A - B)$, this interference will be constructive (maxima of the sinusoidal term) or destructive (minima of the sinusoidal term), as can be seen from Eq. 3.1.

3.2 Plasmonic Interferometers

A plasmonic interferometer can involve different kind of waves, and can be analyzed in the far field or in the near field. Only in the near field we have direct access to SPPs (evanescent waves), and a SPP can interfere with the incident light [29], or with another SPP [30, 31]. At the far field, on the other hand, we work with radiative (non evanescent) waves, i.e. with light that has decoupled from the SPP, but not with the SPP itself. An example of this is the interference between incident light and light that has decoupled from a SPP [32], or between two light beams decoupled for two SPPs

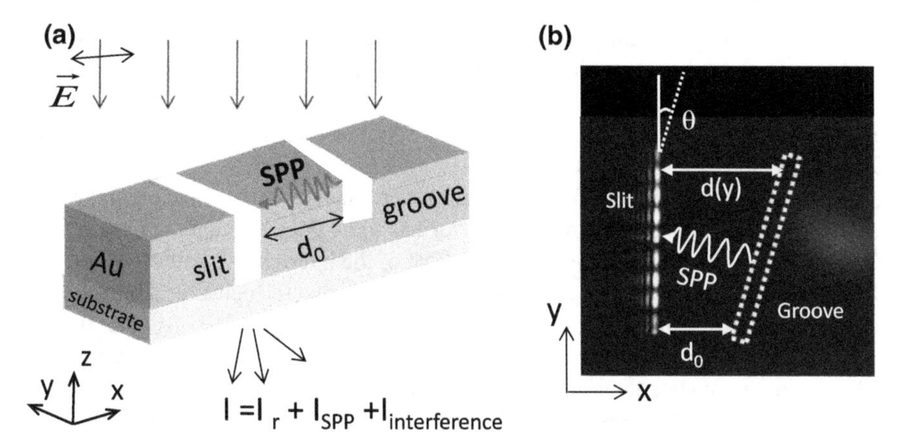

Fig. 3.2 **a** Sketch of the tilted slit-groove plasmonic interferometer. **b** Experimental image of the interference pattern showing some relevant parameters

[33]. There are several configurations for plasmonic interferometers, with parallel slits [11–13, 34], tilted slit-grooves [18, 32], or series of parallel slits or grooves [10, 22].

In this thesis, most of the work will be done in the far-field using a tilted slit-groove interferometer that will be completely described in this chapter. There, interference of incident light with light that has decoupled from a plasmon will take place. However, in Chap. 6 we will also propose different interference configurations for the near field, such as two parallel slits. The plasmonic interferometer that we have used in our far field analysis is the same as in Ref. [32]. It consists of a 200 nm Au layer with a slit and a tilted groove engraved on it (Fig. 3.2). Upon illumination of the interferometer with a p-polarized laser, part of the impinging light goes through the slit and part of it scatters on the groove exciting a SPP that propagates towards the slit. At the slit, the SPP is again scattered and converted back into radiative light. Then, the two beams superimpose, resulting in an interference between light directly transmitted through the slit, $(A_r)^2$, and light coming from the decoupled SPP, $(A_{sp})^2$. The difference of optical paths between the two waves is then the product of the SPP wavevector by the distance traveled by the SPP, which is the distance between slit and groove. Since these are tilted, this distance is going to be different along the slit, $d(y)$. As in the basic interferometer explained above, the intensity collected at the back side of the slit, which we call plasmonic intensity, I, is given by:

$$
\begin{aligned}
E &= A_r + A_{sp}e^{ik_x d} = A_r + A_{sp}e^{ik_x^r d}e^{-k_x^i d}, \\
I &= |E|^2 = A_r{}^2 + A_{sp}{}^2 e^{-2k_x^i d} + 2A_r A_{sp} e^{-k_x^i d}\cos(k_x^r d + \phi_0),
\end{aligned}
\tag{3.2}
$$

being $d = y \cdot tan\theta + d_0$ and $d_0 = d(y = 0)$ where k_x^r, and k_x^i are the real and imaginary parts of k_x, d stands for groove-slit distance, d_0 is the initial separation between groove and slit, θ is the slit-groove angle, y is the position along the slit,

ϕ_0 is an arbitrary phase intrinsic to excitation process of the surface plasmon at the groove, and A_r and A_{sp} are the amplitudes of the directly transmitted light and the excited SPP respectively. This plasmonic intensity exhibits a sinusoidal interference pattern of maxima and minima along the slit axis, since the distance covered by the SPP varies along the slit, $d(y)$.

From this expression (Eq. 3.2), there are several factors that we have to take into consideration. One is that the interferometer has to be thick enough to be opaque to the laser beam, in order to improve the contrast of the interference pattern. However, if it is too thick, we would not be able to see the groove to focus our beam. In our experiments, a 200 nm interferometer has been the right compromise. Another fundamental aspect is that, since metals always have losses, the SPP electromagnetic field is being continuously attenuated during SPP propagation. This has been considered in Eq. 3.2 by explicitly showing the dependence of I with k_x^i. In the interferogram, this would result in an exponential decay in the contrast as $d(y)$ increases. However, this is very small for a single interferometer and as far as the SPP can reach the slit, the effect of the losses is not experimentally observed in the pattern. Nevertheless, the importance of the losses can be experimentally observed when we use different materials. For the same d, a metal with larger absorption leads to a larger reduction of the contrast of the interference than a metal with lower absorption. Indeed this has clearly been seen experimentally when comparing the interferences of a 200 nm Au layer with those of a 200 nm Au/Co/Au trilayer (the one needed for the MP interferometry).

3.3 Magnetoplasmonic Far Field Interferometry

The MP interferometer is a plasmonic interferometer with an appropriate magnetic field applied and a ferromagnetic component so that we have a significant signal modulation. In our case, as it has been exposed in the previous chapter, we have chosen to work with a thin (6 nm) Co layer inside the thick Au layer (Sect. 2.4.2), so that our MP interferometers consist of a 200 nm Au/Co/Au trilayer. The details of the fabrication of the interferometers and their characterization can be found in the Appendix A. As we saw in Sect. 2.4, the appropriate magnetic field to modify the SPP properties is that applied parallel to the sample plane and perpendicular to the SPP propagation direction, i.e. transverse configuration. The left part of Fig. 3.3 shows an sketch of these MP interferometers. When we apply the transverse magnetic field, we induce a change in the SPP wavevector k_x (Eq. 2.14), and as a consequence we modify the SPP optical path in the interferometer ($k_x \cdot d$, as can be seen in Eq. 3.2). This results in a magnetic modulation of the plasmonic intensity I. Actually, this wavevector modulation will produce a shift in the plasmonic interferogram along the slit, Δy, as it is shown in Fig. 3.3. Therefore, at a fixed point within the slit, there will be a modulation on the plasmonic intensity ΔI. Both quantities, ΔI and Δy, can be measured, but it is usually more accurate to detect ΔI (and therefore this is what we have implemented in our experimental configuration).

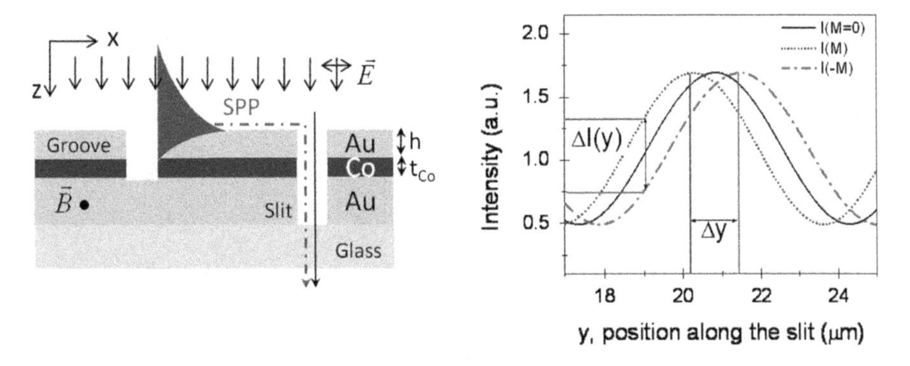

Fig. 3.3 (*Left*) Cross-section of the MP interferometer, where we have introduced a thin ferromagnetic metal in the metallic layer. (*Right*) Illustration of the effect of the magnetic field on the plasmonic interferogram. The magnetic field induces a change in the SPP wavevector and, as a consequence, there is a shift in the interference pattern along the y direction. Then, at a fixed position along the slit y there is a variation in the intensity

When we apply an alternating magnetic field, we can detect the variation of intensity associated with the magnetic field applied, ΔI at each point of the slit. This constitutes the MP interferogram (I_{mp}, Fig. 3.4). From Eqs. 3.2, and 2.14, the expression for the MP intensity can be written as:

$$\begin{aligned} I_{mp} &= \Delta I = I(+M) - I(-M), \\ &= |E(+M)|^2 - |E(-M)|^2 \\ &= -2A_{sp}^2 e^{-2k_x^i d} \sinh(2\Delta k_x^i md) \\ &- 4A_r A_{sp} e^{-k_x^i d}\Big[\sinh(\Delta k_x^i md)\cos(k_x^r d + \phi_0)\cos(\Delta k_x^r md) \\ &+ \cosh(\Delta k_x^i md)\sin(k_x^r d + \phi_0)\sin(\Delta k_x^r md)\Big], \end{aligned} \tag{3.3}$$

where $\Delta k_x = \Delta k_x^r + i \cdot \Delta k_x^i$, M is the magnetization of the sample, and m is the normalized magnetization of the sample (see Eq. 2.14), since we work always in saturation, from now on we will assume that $m = 1$. It is important to notice also that we work at normal incidence, and in that condition the Faraday transverse effect is zero. Therefore, there is only magnetic modulation of the wave that has decoupled from the plasmon, but not of the directly transmitted one.

As Δk_x d is small, we can first assume that $\sinh(\Delta k^i d) \approx \Delta k^i d$, $\cosh(\Delta k^{r,i} d) \approx 1$, $\sin(\Delta k^r d) \approx \Delta k^r d$, and $\cos(\Delta k^r d) \approx 1$, so this expression stays as:

$$\begin{aligned} I_{mp} \approx &- 4A_{sp}^2 e^{-2k_x^i d}\Delta k_x^i d - 4A_{sp}A_r e^{-k_x^i d}\Delta k_x^r d\left[\frac{\Delta k_x^i}{\Delta k_x^r}\cos(k_x^r d + \phi_0)\right. \\ &\left. + \sin(k_x^r d + \phi_0)\right], \end{aligned} \tag{3.4}$$

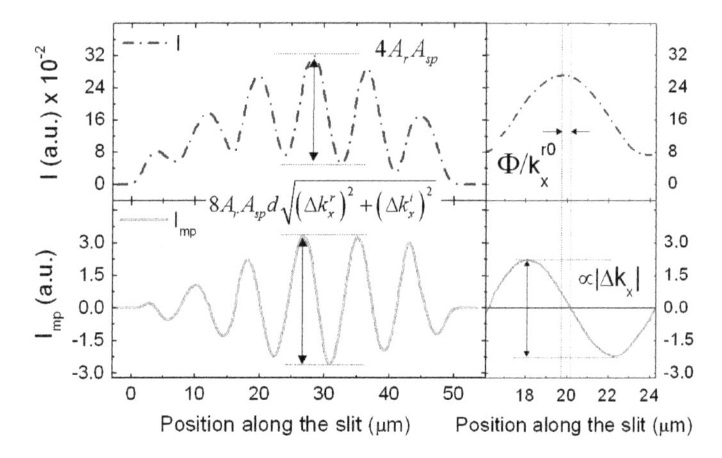

Fig. 3.4 Plasmonic (*upper panel*) and MP(*lower panel*) interferograms as a function of the position along the slit for an interferometer fabricated in a Au/Co/Au trilayer of the following composition: a bottom Au layer of 179 nm, a middle 6 nm Co layer and an upper gold layer of 15 nm. The graphics on the right show a zoom of the ones on the left. The amplitude of the MP signal, related to $|\Delta k_x|$, and the phase shift induced by Δk_x^i, Φ, are indicated

and applying some trigonometry:

$$\begin{aligned} I_{mp} &\approx -4A_{sp}^2 e^{-k_x^i d}\Delta k_x^r d\left[\frac{\Delta k_x^i}{\Delta k_x^r}e^{-k_x^i d} + \frac{A_r}{A_{sp}}\frac{\sin(k_x^r d+\phi_0+\Phi)}{\cos\Phi}\right] \\ &\approx I_{mp\,\text{offset}} - 4\sqrt{(\Delta k_x^r)^2+(\Delta k_x^i)^2}d\,A_{sp}A_r e^{-k_x^i d}\sin(k_x^r d+\phi_0+\Phi), \\ &\text{with } \tan\Phi = \Delta k_x^i/\Delta k_x^r, \end{aligned} \tag{3.5}$$

As it can be seen in Eq. 3.5, except for an offset term, the modulation of the MP signal, is mainly related to the amplitude of the complete wavevector, $|\Delta k_x|$. The modulation of the imaginary part, Δk_x^i, on the other hand, induces a phase shift, Φ, between the plasmonic and the magnetoplasmonic signals (Fig. 3.4). Hence, comparing both the plasmonic (without magnetic field) and the MP interferograms we are able to determine both the module and the imaginary part of the modulation. From the two Eqs. 3.2 and 3.5, it can be observed that the ratio between the contrasts of both signals is proportional to the product $|\Delta k_x|d$. Thus, we can extract the value of the module of the magnetic modulation of the wavevector from our plasmonic and MP interferograms, as it is shown below:

$$\frac{I_{mp\,\text{Contrast}}}{I_{\text{Contrast}}} = \frac{\Delta I_{max} - \Delta I_{min}}{I_{max} - I_{min}} \approx -2|\Delta k_x|d \tag{3.6}$$

Having $\Phi = 0$ ($\Delta k_x^i << \Delta k_x^r$) means that there is no extra phase shift due to the magnetic field, and as a consequence, a maximum in I (cosine dependency) will match a zero of I_{mp} (sine dependency). Figure 3.4 shows how, as Φ is small, the maxima in I almost match the zeros of I_{mp}. In order to get information about Δk_x^i, we first obtain $|\Delta k_x|$ values as it is explained above, then we compare again the interferograms, and with the module $|\Delta k_x|$ values we are able to finally calculate $\Delta k_x{}^i$ from the dephase Φ.

3.4 Experimental Implementation

Once the bases of our MP interferometer have been described, we will explain the experimental setup that we have installed in our laboratory to measure I and I_{mp} and to obtain the value of the magnetic modulation of the SPP wavevector Δk_x. A photograph and a diagram of the actual optical setup is shown in Fig. 3.5. As has been mentioned in Sect. 2.4.2, the MP interferometers that have been implemented [27] consist of Au/Co/Au trilayers. The total thickness of the trilayer is 200 nm, with a 6 nm Co layer placed at different positions, from 5 nm to the surface up to 45 nm. There, a slit of 100 nm width and a tilted groove of 200 nm, both of 50 µm length (the actual interferometer, see Fig. 3.3) have been patterned with a Focused Ion Beam. It has to be said that the slit and groove widths have not been optimized to launch the SPP at any given wavelength, so its efficiency will depend on the particular wavelength used [35]. We have worked with different initial slit-groove separations, $d_0 = 0$, 10, and 20 µm, and two different angles ($\theta = 3°$ and $5°$). All the details of the sample fabrication, as well as the optical, magneto-optical and magnetic characterization of the samples, needed to know their optical and MO dielectric constants and the saturation magnetization, are collected in Appendix A.

3.4.1 Optical Setup

We illuminate our system with a p-polarized laser at different wavelengths: 532, 633, 680, 785, 860, or 890 nm. In order to focus our beam into the interferometer, a $\times 5$ objective is used. Then, the light transmitted to the other side of the slit is collected by a $\times 20$ objective, which produces an intermediate image of the interferometer. At this intermediate image, we place a macroscopic slit that acts as a pinhole, letting pass just a small portion of the whole interference pattern. This pinhole, which can be opened and closed in order to collect the desired portion of the interference pattern, is mounted on a stage controlled by a stepper motor, programed to scan the complete slit of the interferometer in small steps. After that, the beam path is split into two by a beamsplitter. One path focuses again the back side of the slit to a camera connected to the computer, to have a reference of the complete interference pattern, the step size, placement and aperture of the pinhole, and the position of the laser on the interferometer. The other path goes through a cylindrical lens and focuses the beam in a photodiode, which collects the transmitted intensity.

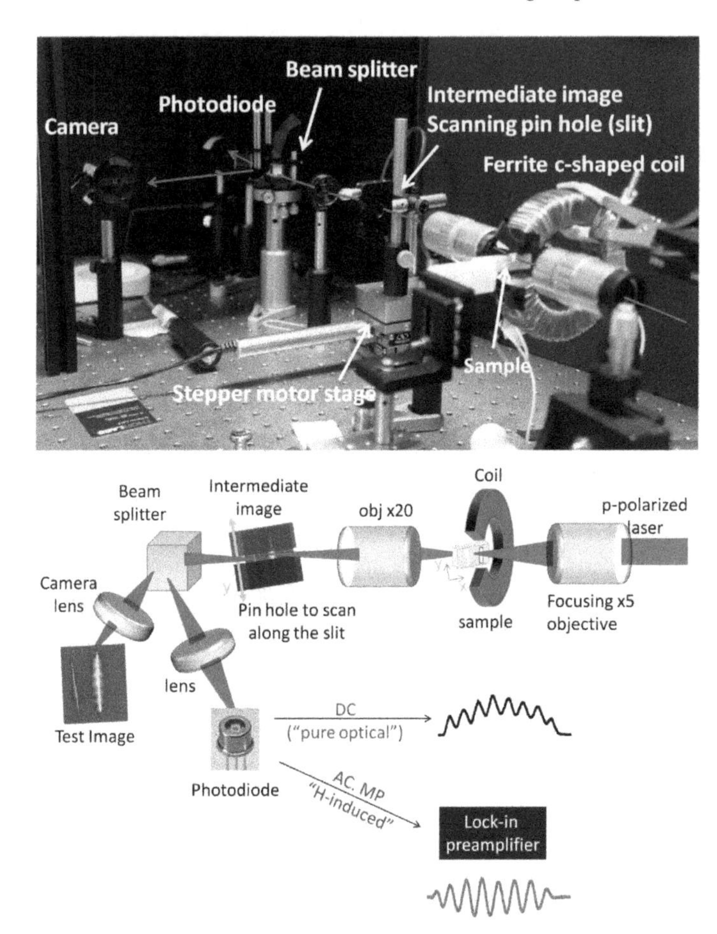

Fig. 3.5 Photograph (*upper image*) and diagram (*lower image*) of the setup for the acMP interferometric measurements. We focus the laser with an objective on the interferometers, that are located inside the gap of a C-shaped coil. With a ×20 objective we form an intermediate image where we place the scanning slit, connected to a stepper motor. From that point we split the beam in two: one arm goes to a camera to have a reference image of the interferometers and the scanning slit position and the other arm is focused using a cylindrical lens into a photodiode

There are some delicate points in this setup. To have well defined interferences, it is fundamental where we focus the laser at the interferometer. The incident spot must be between the slit and the groove but closer to the groove, so that we generate enough SPPs, but also let light pass through the slit, in order to have a good interference contrast (good balance between A_r and A_{sp} in Eq. 3.2). Another trade-off has to be achieved with the aperture of the external slit or pinhole. It must be wide enough to get sufficient signal, but not too opened to prevent excessive convolution of the sinusoidal signal that can smooth down the interferences.

The magnetic field is applied along the y direction, i.e. parallel to the interferometer slit and thus in a transverse configuration. It is applied using a C-shaped coil with the sample placed at its gap. This coil consists of a Cu wiring surrounding a ferrite core of toroidal shape with a gap in it. Since we wanted to work at high frequencies in order to avoid noise from the 50–100 Hz signal of the electric installation, and vibrations, we made an electrical RLC circuit consisting of a resistor, an inductor, and a capacitor. We connect our C-shaped coil of $L \approx 2.2$ mH to a 11.3 μF capacitor, which results in a circuit with a resonance frequency of about 1 kHz. This coil is fed with an alternating current of approximately 2.5 A amplitude at the resonance frequency of the system provided by a Kepco source. This leads to a magnetic field at the coil gap of about 20 mT, which is enough to saturate our interferometers (see Appendix A).

The introduction of the magnetic field is also a sensitive aspect for the experiment. The small distance between the two objectives needed to focus the laser and image the sample made the choice of an adequate manner of applying this magnetic field on the sample essential. In fact, with an open shaped ferrite coil as the one shown in Fig. 3.6b, either our interferometer or its image often moved synchronously with the magnetic field, introducing a noise into our measurements. This was because the magnetic field lines reached the metallic/magnetic objectives or the sample stage. It was solved using a C-shaped (almost closed shape) ferrite core. In this case, the magnetic field is quite localized and does not disturb or moves the surrounding objects (see Fig. 3.6a).

3.4.2 Data Acquisition and Analysis

From the photodiode, we obtain an intensity signal that consists of two components: a continuous (DC) component, that corresponds to I; and an alternating (AC) one at the frequency of the magnetic field, 1 kHz, which is $\Delta I = I_{mp}$. In order to separate and analyze both I and I_{mp}, we carry the signal to an amplifier that converts it from current into voltage and amplifies it. This voltage signal is split then into two

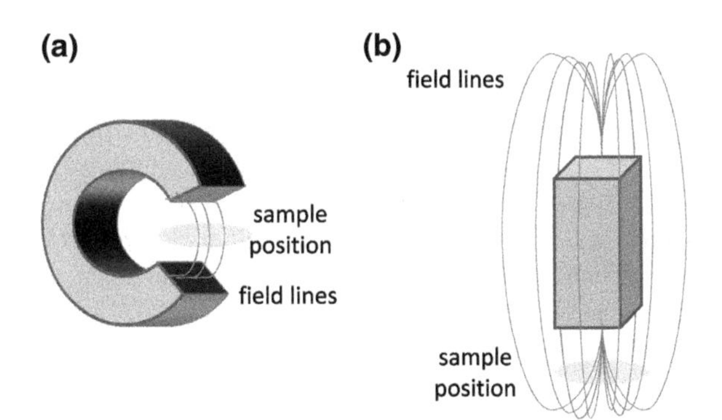

Fig. 3.6 **a** C-shaped coil configuration: the field is concentrated in the gap where the sample is located. **b** Open shaped ferrite coil: the magnetic field leaks out of the sample position, and it can induce some movement of the sample or the objectives

channels. One channel is recorded by a DAQ-card at the computer, which integrates it over a time longer than one cycle and then corresponds to the DC component, the plasmonic intensity. The other channel is fed to a lock-in amplifier, where it is filtered to select only the part of the signal that varies with the frequency of the magnetic field and then results in the AC or I_{mp} component. The lock-in provides both the X and Y components of the AC component:

$$\begin{aligned} X &= I_{mp} \cos(\chi), \\ Y &= I_{mp} \sin(\chi), \\ \left|I_{mp}\right| &= (X^2 + Y^2), \end{aligned} \tag{3.7}$$

where β is the relative phase between the input signal and a reference signal fed to the lock-in at the same frequency as the applied magnetic field. This phase is usually set to zero at the beginning of each measurement. The X and Y components of the AC signal are brought to other channels of the DAQ-card and to the computer. The three signals (DC component, X and Y of the AC component) are registered by a computer, where we have automated the process to calculate and represent all the desired parameters as explained below, and to verify whether there is noise or not, as shown in the next section.

From X and Y of the AC component, both the amplitude and sign of I_{mp} are readily obtained. Although there is an uncertainty in the sign (phase) of the MP signal (we can choose the phase of X or of Y), this is a π jump, which is not relevant for us. Usually one of the components is much bigger (and less noisy) than the other one, so that we pick up the sign of I_{mp} from that one. Summarizing, the plasmonic intensity I is directly the DC component, and the magnetoplasmonic intensity I_{mp} is the module of the total lock-in signal (see Eq. 3.7) with the phase taken from the X or Y signal.

To obtain the module of the modulation we apply Eq. 3.6. As we compare the contrasts of both I and I_{mp}, an offset or background in the signal does not influence the result. Since both are sinusoidal signals, we get I_{max} and I_{min}, and $I_{mp\,max}$ and $I_{mp\,min}$, and then we get the ratio so that:

$$\frac{I_{mp\,max} - I_{mp\,min}}{I_{max} - I_{min}} \approx 2|\Delta k_x|d \tag{3.8}$$

As the ratio of contrast is proportional to the product $|\Delta k_x|d$, and d, the distance between slit and groove is known for a given interferometer, we can extract the value of the module of the magnetic modulation of the wavevector.

On the other hand, to get information about Φ, we compare both interferograms, using Eqs. 3.2 and 3.5. From the original DC signal we subtract the Gaussian shape background (see Fig. 3.4), coming from the intensity distribution of the incident Gaussian beam. Then, we fit this redressed intensity to a sinusoidal expression to obtain the period and phase of the signal. With this data, we fit the MO signal to another sinusoidal dependence and we obtain the phase of this last signal too. With these two phases we are able to figure out the extra phase shiftΦ. Once we have $|\Delta k|$ and Φ, we can extract both Δk_x^r and Δk_x^i by solving the system of two equations with two unknowns. For the cases where Δk_x^i is small, and thus Φ is small, we can initially disregard it and consider that the ratio of the contrasts is proportional to $\Delta k_x^r d$. Then, we obtain Δk_x^r directly from there, and Δk_x^i is obtained from the definition of Φ. It has to be mentioned here that the sinusoidal fittings, and therefore the obtained phase shifts, are very variable and difficult to calculate. As a consequence, large number of measurements are needed for each interferometer in order to have good statistics.

3.4.3 Noise Sources

- Electromagnetic noise

Sometimes an electromagnetic noise whose origin we have been unable to determine appears during the measurements. As we could not identify the cause of this noise, we could not eliminate it, but we developed an easy check to filter the measurements affected by the noise. When we begin a measurement (the registration of an interferogram), we first set the lock-in phase to zero. Our MP intensity responds with a phase shift, β (Eq. 3.7), constant along the measurement. Nevertheless, sometimes this phase β becomes not constant along the measurement. This is related to the presence of the electromagnetic noise, which is dephased with respect to both the lock-in reference signal and the MP intensity, thus the phase β of the recorded signal fluctuates along the measurement. Therefore this noise can be clearly identified when representing X–Y plots of the MO signal from the lock-in. When this noise is not present during the measurement, the X–Y plot is a line, while when the electromagnetic noise is present, the X–Y plot becomes an ellipse.

- Movement of the interferometers

Although the use of C-shaped coils greatly reduces the leakage of electromagnetic field outside the sample position, and therefore the movement of the sample or of the imaging system induced by the magnetic field, this cannot be completely eliminated. However, the influence of this movement can be easily calculated. If we neglect the modulation of the imaginary part, we get an easier expression that still contains all the relevant aspects to take into account:

$$\Delta I = -4A_r A_{sp} \cdot (k\Delta d + \Delta k d) \cdot \sin(kd + \phi_0), \tag{3.9}$$

defining $\Delta d = \frac{d(+M)-d(-M)}{2}$. As it can be seen, in principle, the effect of the movement is adding a factor that is proportional to the movement induced by the magnetic field (Δd). Then, the ratio between I contrast and I_{mp} contrast would be $\Delta kd + k\Delta d$, not only Δkd. Therefore, if the ratio between I_{mp} and I is proportional to the distance d between the slit and the groove, we can trust that the movement is much smaller than the induced modulation and it can be neglected. Indeed, this is what happens in our experiments since we changed to the C-shaped coil.

3.5 Theoretical Simulations

We have already described how we obtain experimentally the magnetic field driven modulation of the SPP wavevector. To further check our results and make new predictions, we have compared our experimental results with the corresponding simulations. Theoretically, our multilayered air/Au/Co/Au/glass system is modeled by means of a scattering matrix formalism where the magneto-optical activity is accounted for by describing the Co layer with the corresponding dielectric tensor [36–40]. Then, a numerical solver for guided modes is applied to find the SPP supported by the metal-air interface in presence and absence of an applied magnetic field **B**. In order to make the simulations as accurate as possible, the optical and MO constants of the different materials involved have been obtained experimentally, and are shown in Appendix A.2.

The scattering matrix formalism consists of relating the waves that come into the system with the ones that get out the system by means of the scattering-matrix, which is the matrix that represents our multilayered system with the corresponding magnetooptical constants (Fig. 3.7). The calculation of this matrix involves Maxwell's equations and the boundary conditions evaluation at each interface. Regarding the calculation of the SPP dispersion relation, which is the parameter that we use the most, it is assumed that no incoming waves are allowed, as well as that it is an exponentially decaying transverse wave, therefore the SPP wavevector is obtained by setting the scattering matrix determinant to zero. The description of this procedure can be seen in more detail at Ref. [36]. This method has provided us the SPP wavevector for our actual interferometer as well as its magnetic modulation for all the figures shown in this thesis. Moreover, we are also capable of obtaining the evolution of the electromagnetic field of the SPP as a function of the vertical distance (as in Fig. 2.4) for different wavelengths using the same formalism [36].

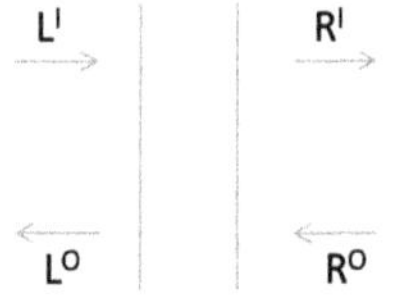

L^I R^I L^O R^O

$$\begin{pmatrix} L^0 \\ R^I \end{pmatrix} = S \begin{pmatrix} R^0 \\ L^I \end{pmatrix}$$

Fig. 3.7 Scattering matrix method. The scattering matrix relates the outcoming waves (no matter whether they are at the *left* or at the *right side*) with the incoming ones. If there is no incoming waves, the only possibility to obtain outcoming ones is when the determinant of the scattering matrix is zero, which leads to the obtainment of the modes of the system

3.6 Magnetoplasmonic Interferometer as a Device

During all this chapter, we have explained the use of MP interferometers as a tool to measure the modulation of the SPP wavevector induced by the magnetic field Δk_x. Nevertheless, as mentioned in Chap. 2, it can be a device per se. In Chap. 5 I will show that it has possibilities as a sensor [41], but this section is devoted to its use as an optical modulator or as an optical switch [27], which is quite promising. Indeed, the magnetic field needed to saturate the device can be easily achieved and it has a quite high potential switching speed, as it has already been mentioned. Besides, the fabrication process of the trilayer is not a complex one and it can be compatible with any plasmonic or photonic circuit and integration process.

Thinking on the MP interferometer as a switch, we modulate the plasmonic intensity I, obtaining I_{mp}, which is the relevant parameter. The first analysis of the performance of these MP interferometeres as modulators considered the parameter I_{mp}/I as a figure of merit. This preliminary implementation, demonstrated in Ref. [27], has modulation depths of about 2 % (as it is shown in Fig. 2.11), and in this thesis we are going to study how that modulation can be further optimized. In Eq. 3.6, it can be seen the expression for the normalized magnetoplasmonic intensity I_{mp}/I, which is proportional to $2|\Delta k_x| \cdot d$. Then, it is obvious that the larger the distance between slit and groove, d, the higher the modulation. However, the separation between the slit and the groove d cannot be increased without limitations, but it must be related to the propagation distance of the plasmon L_{sp}, as this is finite (see Sect. 2.1.1). Thus, a more appropriate figure of merit (FOM) to analyze the performance of the MP interferometer is the product $2\Delta k_x \cdot L_{sp}$. We have thus used this figure of merit during this thesis when dealing with the optimization of the system for the design of plasmonic modulators. Moreover, we have seen experimentally that with distances between the slit and the groove of $d = 3L_{sp}$ we have a reasonable contrast, which means that this figure of merit can be increased up to $6|\Delta k_x| \cdot L_{sp}$. In Fig. 3.8 it is presented both Δk_x and the figure of merit of an Au/Co/Au interferom-

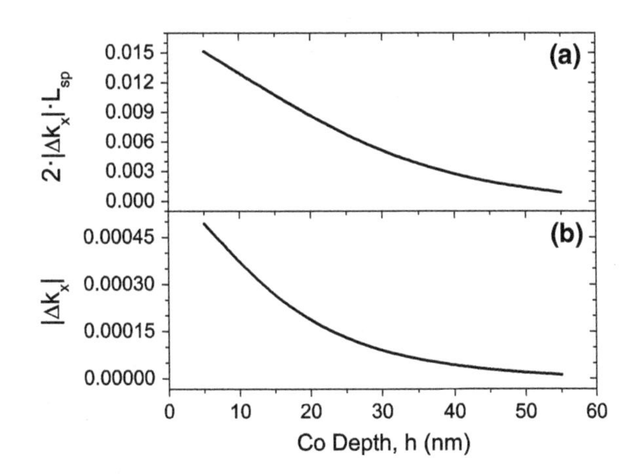

Fig. 3.8 **a** Magnetic modulation of the SPP wavevector and **b** figure of merit as a function of the Co depth at 785 nm. h is the thickness of the upper Au layer of the interferometer

eter as a function of the Co depth. It can be then seen that there are two fundamental parameters regarding the materials used for MP interferometry: the magnitude of the magnetic effect on the SPP wavevector (Δk_x), and the absorption of the material, that will affect both L_{sp} and the contrast observed in the interferences. As it can be seen from Fig. 3.8, although the wavevector modulation decreases exponentially with the position of the Co layer, the increase of the SPP propagation distance due to the reduction of absorption, smooths the decay of the figure of merit. Therefore, and in order to prevent oxidation of the Co layer, trilayer systems of $h = 10 - 15$ nm would be quite adequate.

References

1. A. Labeyrie, Astrophys. J. **196**, L71–L75 (1975)
2. J.P. Berger, P. Haguenauer, P. Kern, K. Perraut, F. Malbet, I. Schanen, M. Severi, R. Millan-Gabet, W. Traub, Astron. Astrophys. **376**, 31–34 (2001)
3. C. Dorrer, B. de Beauvoir, C. Le Blanc, S. Ranc, J.-P. Rousseau, P. Rousseau, J.-P. Chambaret, F. Salin, Opt. Lett. **24**, 1644–1646 (1999)
4. T.J. McIntyre, C. Maurer, S. Bernet, M. Ritsch-Marte, Opt. Lett. **34**, 2988–2990 (2009)
5. A. Monmayrant, M. Joffre, T. Oksenhendler, R. Herzog, D. Kaplan, P. Tournois, Opt. Lett. **28**, 278–280 (2003)
6. A. Kussrow, C.S. Enders, D. Bornhop, J. Anal. Chem. **84**, 779–792 (2012)
7. Y. Jung, S. Lee, B.H. Lee, K. Oh, Opt. Lett. **33**, 2934–2936 (2008)
8. Y.-D. Wu, T.-T. Shih, M.-H. Chen, Opt. Express **16**, 248–257 (2008)
9. A.A. Michelson, E. Morley, Am. J. Sci. **34**, 333–345 (1887)
10. C.-H. Gan, G. Gbur, Plasmonics **3**, 111–117 (2008)
11. G. Gay, O. Alloschery, B. Viaris de Lesegno, C. O'Dwyer, J. Weiner, H.J. Lezec, Nature **2**, 262–267 (2006)
12. D. Pacifici, H.J. Lezec, H.A. Atwater, Nat. Photonics **1**, 402–406 (2007)
13. V.V. Temnov, U. Woggon, J. Dintinger, E. Devaux, T.W. Ebbesen, Opt. Lett. **32**, 1235–1237 (2007)
14. E. Verhagen, J.A. Dionne, L.K. Kuipers, H.A. Atwater, A. Polman, Nano Lett. **9**, 2925 (2008)
15. M. Cohen, Z. Zalevsky, R. Shavit, Nanoscale **5**, 5442–5449 (2013)
16. X. Wu, J. Zhang, J. Chen, C. Zhao, Q. Gong, Opt. Lett. **34**, 392–394 (2009)
17. Y. Gao, Q. Gan, Z. Xin, X. Cheng, F.J. Bartoli, ACS Nano **5**, 9836–9844 (2011)
18. X. Li, Q. Tan, B. Bai, G. Jin, Opt. Express **19**, 20691–20703 (2011)
19. J. Feng, V.S. Siu, A. Roelke, V. Mehta, S.Y. Rhieu, G. Palmore, R. Tayhas, D. Pacifici, Nano Lett. **12**, 602–609 (2012)
20. O. Yavas, C. Kocabas, Opt. Lett. **37**, 3396–3398 (2012)
21. T. Bian, B.-Z. Dong, Y. Zhang, Plasmonics **8**, 741–744 (2013)
22. Y. Gao, Z. Xin, Q. Gan, X. Cheng, F.J. Bartoli, Opt. Express **21**, 5859–5871 (2013)
23. S.I. Bozhevolnyi, V.S. Volkov, E. Devaux, J.-Y. Laluet, T.W. Ebbesen, Nature **440**, 508–511 (2006)
24. X. Li, W. Li, X. Guo, J. Lou, L. Tong, Opt. Express **21**, 15698–15705 (2013)
25. M.J. Dicken, L.A. Sweatlock, D. Pacifici, H.J. Lezec, K. Bhattacharya, H.A. Atwater, Nano Lett. **8**, 4048–4052 (2008)
26. J. Gosciniak, S.I. Bozhevolnyi, T.B. Andersen, V.S. Volkov, J. Kjelstrup-Hansen, L. Markey, A. Dereux, Opt. Express **18**, 1207–1216 (2010)
27. V.V. Temnov, G. Armelles, U. Woggon, D. Guzatov, A. Cebollada, A. Garcia-Martin, J.M. Garcia-Martin, T. Thomay, A. Leitenstorfer, R. Bratschitsch, Nat. Photonics **4**, 107–111 (2010)

28. P. Hariharan, Rep. Prog. Phys. **54**, 339 (1991)
29. B. Wang, L. Aigouy, E. Bourhis, J. Gierak, J.P. Hugonin, P. Lalanne, Appl. Phys. Lett. **94**, 011114 (2009)
30. L. Aigouy, P. Lalanne, J.P. Hugonin, G. Julié, V. Mathet, M. Mortier, Phys. Rev. Lett. **98**, 153902 (2007)
31. J. Chen, M. Badioli, P. Alonso-Gonzalez, S. Thongrattanasiri, F. Huth, J. Osmond, M. Spasenovic, A. Centeno, A. Pesquera, P. Godignon, A. Zurutuza-Elorza, N. Camara, F.J. Garcia de Abajo, R. Hillenbrand, F.H.L. Koppens, Nature **487**, 77–81 (2012)
32. V.V. Temnov, K. Nelson, G. Armelles, A. Cebollada, T. Thomay, A. Leitenstorfer, R. Bratschitsch, Opt. Express **17**, 8423–8432 (2009)
33. A. Drezet, A.L. Stepanov, A. Hohenau, B. Steinberger, N. Galler, H. Ditlbacher, A. Leitner, F.R. Aussenegg, J.R. Krenn, M.U. Gonzalez, J.C. Weeber, Europhys. Lett. **74**, 693–698 (2006)
34. H.F. Schouten, N. Kuzmin, G. Dubois, T.D. Visser, G. Gbur, P.F.A. Alkemade, H. Blok, G.W.T. Hooft, D. Lenstra, E.R. Eliel, Phys. Rev. Lett. **94**, 053901 (2005)
35. J. Renger, S. Grafström, L.M. Eng, Phys. Rev. B **76**, 045431 (2007)
36. J.B. Gonzalez-Diaz, *MagnetoPlasmonics. MagnetoOptics in Plasmonics Systems*. Ph.D. thesis, Universidad Autonoma de Madrid, Spain, 2010
37. J.B. Gonzalez-Diaz, A. Garcia-Martin, G. Armelles, J.M. Garcia-Martin, C. Clavero, A. Cebollada, R.A. Lukaszew, J.R. Skuza, D.P. Kumah, R. Clarke, Phys. Rev. B **76**, 153402 (2007)
38. E. Ferreiro-Vila, J.B. Gonzalez-Diaz, R. Fermento, M.U. Gonzalez, A. Garcia-Martin, J.M. Garcia-Martin, A. Cebollada, G. Armelles, D. Meneses-Rodriguez, E. Muoz Sandoval, Phys. Rev. B **80**, 125132 (2009)
39. K. Postava, J. Pistora, S. Visnovsky, Czech J. Phys. **49**, 1185–1204 (1999)
40. J.F. Torrado, J.B. Gonzalez-Diaz, A. Garcia-Martin, G. Armelles, New J. Phys. **15**, 075025 (2013)
41. D. Martin-Becerra, G. Armelles, M.U. Gonzalez, A. Garcia-Martin, New J. Phys. **15**, 085021 (2013)

Chapter 4
Magnetic Modulation of SPP in Au/Co/Au Trilayers

Some of the main results are shown here. The mechanisms involved in the plasmonic modulation by a magnetic field are explained, as well as an optimization of the device is proposed. The spectral dependence of the SPP wavevector modulation is analyzed experimentally and some different materials and conditions to optimize the response are proposed. The parameters that determine the spectral dependence and the influence of the electromagnetic field distribution along the interface on this modulation are shown. Finally, it is demonstrated that the modulation can be increased by adding a dielectric overlayer on top of the interferometer.

In Chap. 2, it is explained how it is possible to modify the SPP wavevector by applying a magnetic field. It is also justified that an appropriate system for this is a noble metal/ferromagnetic metal/noble metal trilayer. Moreover, in Chap. 3, it is shown how this concept can be implemented in a MP interferometer. There it is detailed how, with the MP interferometer, we can extract information about the magnetic modulation of the SPP wavevector, but also how it is possible, with a magnetic field, to modulate the transmitted intensity. In a previous work, the proof of concept of these magnetoplasmonic interferometers has been demonstrated [1]. The obtained intensity modulation achieved there is of about 2 %, being the work made at a single wavelength and without any optimization, showing only the real part of Δk_x. However, the practical application of a magneto plasmonic interferometer as an optical switch requires further optimization of the multilayer films to achieve the maximum possible SPP wavevector modulation. In this chapter we will carry out a deep experimental and theoretical analysis of magnetic modulated systems with these MP interferometers in order to understand the physical parameters governing the modulation, and to optimize the performance of magnetoplasmonic interferometers as devices. In particular, the spectral dependence of the SPP wavevector modulation will be studied. This has allowed us to determine the role of the SPP field distribution in the final value of the modulation as well as to identify the optimum spectral region to use these interferometers. Furthermore, we will finish by analyzing a very simple way to increase its magnetic modulation by covering the interferometers with a thin dielectric overlayer.

D. Martín Becerra, *Active Plasmonic Devices*, Springer Theses,
DOI 10.1007/978-3-319-48411-2_4

4.1 Spectral Dependence of the SPP Wavevector Magnetic Modulation

In this section, we will study the spectral evolution of the modulation of the SPP wavevector in the visible and in near-infrared (from 500 nm to 1 μm) for Au/Co/Au multilayered systems. This way, we will find the optimum spectral range for application purposes. The magnetic field allows the modulation of both the real and imaginary parts of the SPP wavevector. Thus we will also measure the magnetic modulation of the imaginary part of the SPP, which has not been done before, and analyze its physical meaning. As it will be shown, the relative weight of each part depends on the wavelength range, so this aspect has to be taken into account if one wants to optimize the response in these systems. We also discuss the dependence of both modulations on the different optical and magneto-optical parameters of the system as a way to understand the main parameters that determine their value. Finally, we will expand the study to other ferromagnetic metals.

4.1.1 *Evolution of Δk_x^r with the Position of the Co Layer for Different Wavelengths*

To provide a preliminary insight on the parameters governing the values of the modulation of the SPP wavevector (Δk_x), we show here the evolution of the real part of Δk_x with the position of the Co layer (h) for different wavelengths (633, 690, 785, 860, and 980 nm), both theoretical and experimentally. The interferometers considered are those described in Sect. 3.4 and sketched in Fig. 3.2. The values of Δk_x have been obtained with the procedure explained in Sect. 3.3, and the numerical simulations were performed as indicated in Sect. 3.5. The obtained results are presented in Fig. 4.1, where it can be seen that, for a given wavelength, there is always an exponential decay of the modulation of the real part of the SPP wavevector as a function of the depth of the Co layer. As it was shown in Sect. 2.4.2, this is due to the evanescent nature of the SPP: its electromagnetic field decays exponentially inside the metal, and therefore the deeper the Co (ferromagnetic) layer the less electromagnetic field it experiences, and as a consequence the less magnetic modulation will be achieved [1]. So, this shows that the distribution of the SPP electromagnetic field is an important factor to determine the absolute value of the wavevector modulation. Moreover, the slope of the exponential decay increases with the wavelength (although the difference decreases for larger wavelengths). This is due to the difference in the penetration depths of the electromagnetic field in the trilayer, δ, for the different wavelengths. As discussed in Sect. 2.1.1 (see Fig. 2.3), the penetration depth in the metal decreases when the wavelength increases because of the stronger screening of the metal to the electromagnetic field at lower frequencies. Our calculated values agree well with those otained in Ref. [2], where the penetration depth at 808 nm was 13 nm. Finally, the data presented in Fig. 4.1 also show that, for the same position of the Co layer, the values of $\Delta k_x{}^r$ decrease for increasing wavelengths, and the

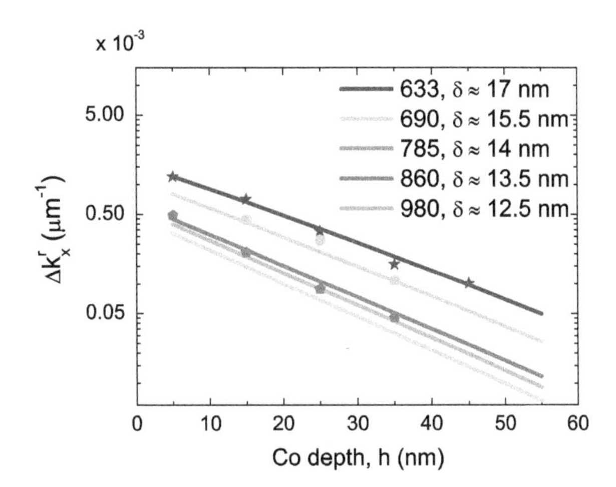

Fig. 4.1 Experimental data (*filled symbols*, only at 633, 690, and 785 nm) and simulations (*lines*) of the evolution of Δk_x^r as a function of the Co layer position for different wavelengths in a logarithmic scale. The calculated values of the penetration distance (skin depth) for each wavelength are also shown

modification of δ is not enough to account for these differences. In the next subsection we will analyze this spectral evolution in more detail.

4.1.2 Spectral Behavior of Δk_x^r and Δk_x^i

To analyze the spectral behaviour of Δk_x^r and Δk_x^i, we will focus on a single value of the position of the Co layer, $h = 15$ nm. We have measured Δk_x^r and Δk_x^i by illuminating the magnetoplasmonic interferometers with several lasers of different wavelengths ($\lambda_0 = 532$, 633, 680, 785, 860 and 980 nm). The obtained results are plotted in Fig. 4.2 (filled symbols). The upper graph corresponds to the real part of the modulation, and the lower graph to the imaginary part. Together with the experimental results, Fig. 4.2 also includes, as dashed lines, the values of Δk_x^r and Δk_x^i obtained from numerical simulations. As it can be seen, the agreement between the experimental and the simulations is very good.

As Fig. 4.2 shows, the general trend for both real and imaginary parts of the wavevector modulation is that the modulation value decreases with wavelength. In principle, this is an unexpected behavior as the absolute value of the MO constants of Co increase with wavelength (see Fig. 4.3a and Appendix A, Figure A.9), and its origin will be discussed in detail in the next section. The same behavior has been obtained for three different positions of the Co layer [2], as can be deduced from Fig. 4.1. Regarding the real part of Δk_x, it presents a peak at low wavelengths ($\lambda_0 = 530$ nm).[1] As it has been mentioned in Sect. 2.1.1, this peak is not associated

[1] Although this peak has not been experimentally confirmed since the very short plasmon propagation distance at wavelengths below 530 nm has prevented the measurement, the good agreement between measurements and simulations in the rest of the spectral range allows us to trust the presence of this peak.

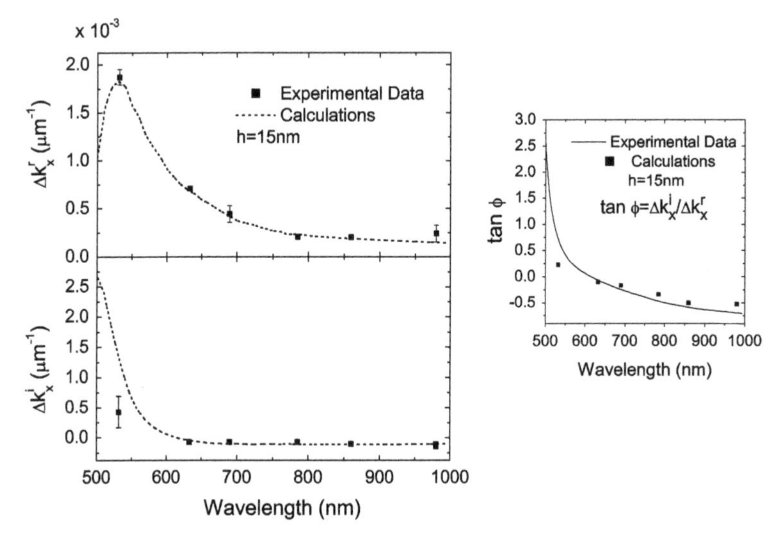

Fig. 4.2 (*Left*) Wavelength dependence of Δk_x^r (*upper graph*) and Δk_x^i (i) obtained for magnetoplasmonic trilayers with $h = 15$ nm. The *filled symbols* correspond to experimental values and the *dashed lines* to values obtained numerically. (*Right*) Calculated and experimental ratio $\Delta k_x^i/\Delta k_x^r$, which corresponds to $\tan\phi$

to a resonant behavior but to the presence of absorption in the metal, that makes the SPP dispersion relation "bend" around this wavelength close to the surface plasmon resonance frequency [3]. This will be discussed in more detail in the next subsection. As for the imaginary part, it is, in general, smaller than the real part in most of the spectral range. Only in the small wavelength region ($\lambda_0 < 520$ nm), again close to the surface plasmon resonance frequency, the imaginary part dominates. At this wavelength range, there is a discrepancy between theory and experiment, mainly due to the very short SPP propagation distance, which worsened significantly the signal-to-noise ratio during experiment. Taking into account the ratio between imaginary and real part modulations, shown at the right graph included in Fig. 4.2, three spectral regions can be defined. For lower wavelengths ($\lambda_0 < 520$ nm), the dominant part of the modulation is the imaginary one, but this is a region where plasmon losses are too high to envisage any practical application of these systems (see Fig. 2.2b as a reference). For the central range, between $\lambda_0 = 550$ nm and around 750 nm, the relevant component of the modulation is the real part, being the imaginary one a small perturbation that can be discarded ($\tan\phi \approx 0$) [1]. Finally, for the long-wavelength region ($\lambda_0 > 750$ nm) the ratio increases, so that the imaginary part has to be taken into account in order to accurately describe the system. Taking all this into account, we can see how magnetoplasmonic modulators can be implemented in the form of interferometers, based on the intensity variation given by Eq. 3.5. Although the magnetic field induces Δk_x^i, modulators based on attenuation would not be possible because this term only dominates for a spectral range not relevant. Thus, these magnetoplasmonic modulators will be based on the modulation of the

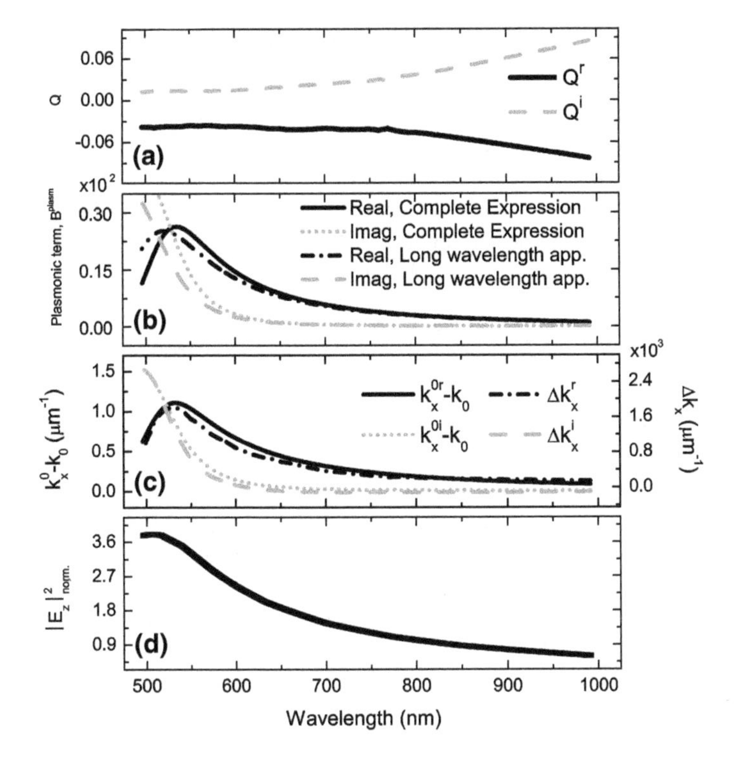

Fig. 4.3 **a** Wavelength dependence of the magneto-optical parameter Q for our Co layers. **b** Evolution of the plasmonic term from the analytical expression of Δk_x for the AuCoAu trilayers. Both long wavelength approximation and complete expressions are plotted. **c** Evolution with wavelength of the separation of the SPP wavevector from the light line (*left axis*) compared with the evolution of Δk_x (*right axis*) for a AuCoAu trilayer (in particular, the case for $h = 15$ nm is plotted). **d** Spectral dependence of the normalized intensity of the SPP electromagnetic field at the middle of the Co layer in a AuCoAu trilayer with $h = 15$ nm

real part of the SPP wavevector, Δk_x^r. In the long-wavelength region, however, the contribution of Δk_x^i (ϕ) cannot be neglected and it is important to find the slit-groove distance of maximum response.

Although the magnetic field induces Δk_x^i, modulators based on attenuation [4–8] would not be possible because this term only dominates for a not relevant spectral range. Thus these magnetoplasmonic modulators will be based on the modulation of the real part of the SPP wavevector [1, 9, 10], Δk_x^r, but in the long-wavelength regime, the contribution of Δk_x^i can not be neglected.

4.1.3 Different Factors Governing Δk_x

To gain a deeper understanding on the spectral behavior of the modulation of k_x, an equation relating the implied parameters is necessary. An analytical expression for

Δk_x in a noble/ferromagnetic /noble metal trilayer can be found in Sect. 2.4.2. Here we can restrict our analysis to the long-wavelength regime ($\varepsilon_d \ll |\varepsilon_{Au}|$) (Eq. 2.20), as the equations are simpler to write but the involved mechanisms are the same that considering the complete Eq. 2.19. This expression has been obtained by applying Maxwell equations in a system consisting of a trilayer and a semi-infinite dielectric, taking the dielectric tensor to account for the optical and magneto-optical response of the ferromagnetic layer and in the approximation that this ferromagnetic layer is very thin [1, 11]. Then, solving the relation dispersion of the SPP:

$$\Delta k_x \approx -2t_{Co}e^{-2hk_z^{Au}} \times \frac{k_0^2\varepsilon_d^2}{\varepsilon_{Au}} \times Q^{Co} \equiv A^{str} \times B^{plasm} \times Q^{Co} \tag{4.1}$$

We can split Eq. 2.20 (Eq. 2.19 can be split similarly in three terms), as shown in Eq. 4.1: a structural term, A^{str}; a purely optical or plasmonic term, B^{plasm}; and a MO term, Q^{Co}. The structural term A^{str} is governed by geometric parameters regarding the structure (t_{Co} and h) and the z component of the SPP wavevector inside the metal, k_z^{Au}. This last term is associated with the exponential decay of the SPP electromagnetic field intensity inside the metal layer, and strictly speaking would have to be included in B_{plasm} as it contains optical information. However, as in practical realizations of magnetoplasmonic interferometers the value of h will be of the order of $(k_z^{Au}/2)$ or smaller (here for example we analyze $h = 15$ nm), and the dependece of k_z^{Au} with the wavelength is weak, we have introduced it in A^{str}. The plasmonic term B^{plasm} includes all the optical parameters of the system except for those related to the ferromagnetic layer: ε_{Au}, ε_d (1 in the case analyzed here), and k_0. Thus, this term takes into account the properties of the surface plasmon polariton supported by the system for each wavelength. Finally, the MO properties of our ferromagnetic layer appear in the last term through the magneto-optical parameter Q, defined, as shown in Sect. 2.3, as $Q^{Co} = i\frac{\varepsilon_{xz}^{Co}}{\varepsilon_{xx}^{Co}}$ (for ε_{xx}^{Co} and ε_{xz}^{Co} the optical and magneto-optical constants of Co, respectively). The spectral behaviour of Δk_x is therefore determined by Q^{Co} and B^{plasm}. Figure 4.3a, b present the dependence on the wavelength of these two quantities, using the optical and magneto-optical constants experimentally determined for the metallic layers (see Appendix A). Figure 4.3b shows the plasmonic term obtained from the exact expression of Eq. 2.19 as well as the long-wavelength approximation (Eq. 4.1). As it can be seen, the long-wavelength approximation is very good beyond 700 nm.

From the results depicted in Fig. 4.3a, b we can establish that the spectral behavior of Δk_x in our AuCoAu magnetoplasmonic multilayers is dominated by the plasmonic term: the absolute value of Q increases with the wavelength while Δk_x shows the same decreasing behavior as the plasmonic term as well as the presence of a small peak in the real part for small wavelengths. This shows that the evolution of the SPP properties with the wavelength is very important to determine the possible magnetic modulation achieved.

A significant SPP property is the SPP field vertical spreading, which affects the amount of field reaching the ferromagnetic layer. Actually, we have seen in Sect. 4.1.1

that the evolution of the SPP wavevector modulation with h is proportional to the exponential decay of the SPP electromagnetic field inside the metal layer. In the same way, we can invoke here a similar mechanism: the redistribution of the SPP electromagnetic field spreading as a function of the wavelength is the main SPP property that determines the spectral evolution of the magnetic modulation of the SPP wavevector (Δk_x). As we saw in Sect. 2.1, the information of the spectral dependence of SPP electromagnetic field confinement, implicit in B^{plasm}, is contained in the dispersion relation. If we consider the separation of the SPP wavevector from the light line, $k_x^0 - k_0$, the bigger separation occurring at shorter wavelengths implies a stronger evanescent decay and therefore a higher confinement. As a consequence, the magnitude $k_x^0 - k_0$ is also able to describe the spectral shape of Δk_x, and we have found that it does it in a very accurate way for both Δk_x^r and Δk_x^i for our AuCoAu trilayered system, as shown in Fig. 4.3c. This can be understood as follows: for longer wavelengths the SPP wavevector is closer to the light line and thus the associated electromagnetic field is more spread out of the interface, while for smaller wavelengths k_x increases and the SPP field becomes more confined to the interface and the presence of the Co layer is strongly felt (see Figs. 2.3 and 2.4). Our results show that this effect has a bigger influence on Δk_x than the increase of Q^{Co} with wavelength.

In fact, we can rewrite Eq. 2.19 so that the amount of SPP field inside the Co layer appears explicitly. For that, we refer to the SPP magnetic field component intensity, $|E_z|^2$, at the Co layer position. This magnitude is normalized so that for each wavelength the integral of $|E_z|^2$ along the z-axis is equal to 1 (the energy density is the same in all cases). Taking into account that for a Au/dielectric semi-infinite interface the normalized SPP magnetic field can be written as $(2k_z^{Au}k_z^d/(k_z^{Au} + k_z^d))^{1/2}e^{-k_z^{Au}z}$, with k_z^{Au} and k_z^d the z components of the SPP wavevector in the metal and dielectric, respectively, and using some arithmetics, we obtain the following expression:

$$\Delta k_x \approx t_{Co} \times \frac{k_0\varepsilon_d\varepsilon_{Au}\sqrt{-(\varepsilon_d + \varepsilon_{Au})}}{\varepsilon_d^2 - \varepsilon_{Au}^2} |E_z^{Co}|^2 \times Q^{Co}. \tag{4.2}$$

In Fig. 4.3d, we plot $|E_z^{Co}|^2$ as a function of the wavelength. The SPP field does indeed decrease with the wavelength, as Δk_x does, confirming that the amount of the SPP electromagnetic field in the MO layer is a main parameter to take into account in the spectral behavior of Δk_x. The SPP field, however, does not reproduce all the details found in the spectral evolution of Δk_x^r and Δk_x^i. This is due to the influence of other optical parameters, as can be seen from Eq. 4.2. The term containing the proportionality to the SPP field at the Co layer interface contains some other optical factors whose particular spectral response slightly modify the pure decay of the field with the wavelength, to finally define the obtained spectral shape of Δk_x (real and imaginary part). Those optical factors will also be implicit in the $k_x^0 - k_0$ separation, so that this quantity provides a more accurate description of the shape of the evolution of Δk_x with the wavelength.

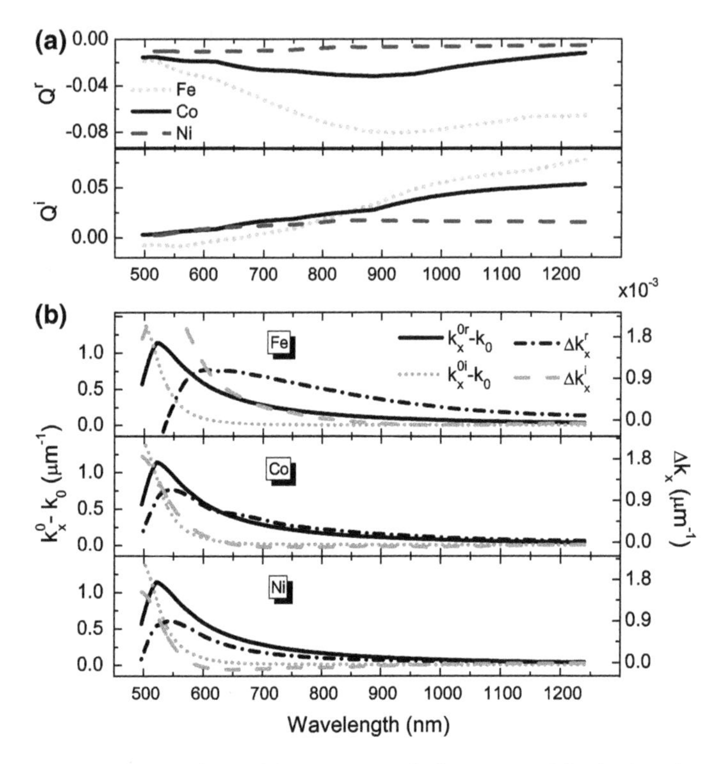

Fig. 4.4 **a** Wavelength dependence of the magneto-optical parameter Q for the three ferromagnetic metals Fe, Co, and Ni, calculated from typical optical and magneto-optical constants for the bulk materials referred in the literature: ε_{xx}^{Fe} [13] ε_{xz}^{Fe}, [14], ε_{xx}^{Co} [13] ε_{xz}^{Co}, [14], ε_{xx}^{Ni} [15], and ε_{xz}^{Ni} [16]. **b** (*Right axis*) Dependence with wavelength of the real part and the imaginary part of calculated Δk_x for trilayers of Au/FerromagneticMetal/Au with $h = 15$ nm and $t_{Ferro} = 6$ nm, being Fe (*top panel*), Co (*middle panel*) and Ni (*bottom panel*) the ferromagnetic metal. (*Left axis*) Comparison of the calculated Δk_x with the separation of the SPP wavevector from the light line

4.1.4 Extension to Other Ferromagnetic Metals

The discussion performed above regarding the spectral evolution of the magnetic field driven modulation of SPP wavevector for the particular AuCoAu trilayer system could be, in principle, extended to the other ferromagnetic metals, Fe and Ni, as the different parameters, k_x properties and Q, evolve in a similar way. The right axis of Fig. 4.4b shows the calculated values of Δk_x^r and Δk_x^i for trilayers of Au/ Ferromagnetic Metal/Au with a ferromagnetic metal layer of 6 nm and $h =$ 15 nm. To make the analysis as much general as possible, instead of employing optical and MO dielectric constants experimentally determined in our laboratory, we have chosen in this case to use well-established values that can be found in the literature for the bulk forms of the analyzed materials: Au—dielectric constant [12]; Fe and Co, optical [13] and MO [14] constants; and Ni, optical [15] and MO [16] constants. In Fig. 4.4a, the corresponding magneto-optical parameters Q for Fe, Co and Ni are shown. As it can be seen in Fig. 4.4b, the greatest absolute values of modulation are obtained for Fe, and Ni provides the lowest values, in agreement with the values of Q parameter for each metal. Regarding the evolution with wavelength, Δk_x behaves in the expected decreasing way, even for increasing values of Q, confirming the predominance of the plasmonic term to determine this evolution. Figure 4.4b also compares the evolution of Δk_x with the distance of k_x^0 to the light line (left axis), with again a good agreement between both quantities. The matching between Δk_x and $k_x^0 - k_0$ trends is slightly worse in the case of Fe, mainly at small wavelengths, because of the stronger dependency of Q^{Fe} with wavelength in this regime, which gives rise to a stronger convolution of the product of the MO and the plasmonic term in Eq. 4.1 for this ferromagnetic metal. Summarizing, all these results shown that, for noble metal/ferromagnetic metal/ noble metal, the behaviour of SPP electromagnetic field is the main parameter that determines the evolution of the modulation of the SPP wavevector. The final obtained value is of course proportional to the Q value of the corresponding ferromagnetic metal.

4.1.5 Spectral Evolution of the Figure of Merit

As explained in Sect. 3.6, the magnetoplasmonic interferometers can act as modulators where the modulated intensity I_{mp} is proportional to the product of $2\Delta k_x \times d$ [1]. Therefore, higher modulations can be obtained by increasing this distance. However, it is well known that the propagation distance of SPPs depends on the wavelength (see Sect. 2.1.1), so the appropriate parameter to evaluate the optimal spectral range for the performance of magnetoplasmonic (MP) modulators is the figure of merit defined in Sect. 3.6, the product $2\sqrt{(\Delta k_x^r)^2 + (\Delta k_x^i)^2} \times L_{sp}$. Figure 4.5 shows this

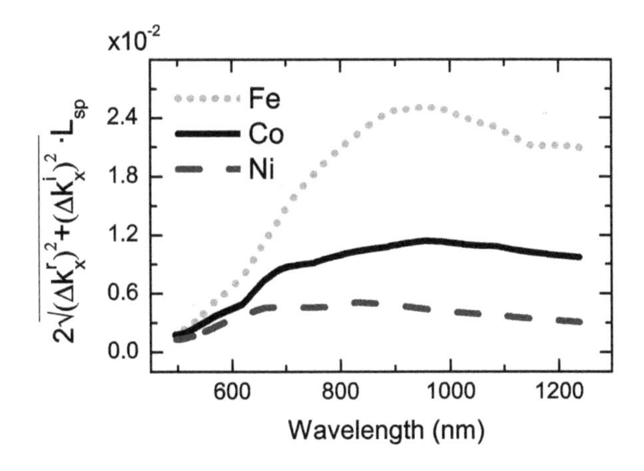

Fig. 4.5 Spectral evolution of the figure of merit $2\sqrt{(\Delta k_x^r)^2 + (\Delta k_x^i)^2} \times L_{sp}$ for the same Au/Ferromagnetic Metal/Au trilayers as in Fig. 4.4, with the ferromagnetic metal being Fe, Co, and Ni

figure of merit for the three Au/Ferromagnetic Metal/Au trilayers with Fe, Co and Ni analyzed in the previous subsection. The best performance in absolute terms is again obtained by Fe, being Ni the worst one, in agreement with the amount of MO response strength as indicated by Q. In the wavelength region with $\Delta k_x^i > \Delta k_x^r$, $\lambda_0 < 520$ nm, the figure of merit is very small since the propagation distance of the SPP is almost zero. In the spectral range where SPPs start having propagation distances of a few microns and therefore could be employed in photonic devices (SPP propagation distance $L_{sp} > 3\,\mu$m at $\lambda_0 > 600$ nm for a Au/Co/Au trilayer with the Co layer placed at $h = 15$ nm), the figure of merit increases and achieves its maximum value at around 1 μm. At these wavelengths, as it was seen from Fig. 4.2, both real and imaginary components of the modulation are relevant. Finally, for wavelengths $\lambda_0 > 1\,\mu$m, the figure of merit decreases a little, seeming to reach a saturation value. Taking the values of Fig. 4.5 into consideration, a magnetoplasmonic modulator consisting of a Au/Fe/Au interferometer with a separation distance of $d = 3L_{sp}$ (achievable from our experimental experience, since we can make $I_{sp} > I_r$ by controlling the relative position of the incident laser spot and the groove) could provide intensity modulations ($\Delta I/I$) of around 7.5 % in the optimal spectral range (950 nm). However, compared with other active plasmonic systems, this is a still limited value in order to develop practical devices, so it will be desirable to be able to increase this value.

4.2 Enhancement of the Magnetic Modulation

We have seen that working in the appropriate wavelength regime and selecting the optimum material, the magnetic modulation of our interferometers can be maximized. However, the achieved values are still slightly small for practical applications. Is there any way to increase this modulation? We will show in this section that there is indeed an easy way to do so, the deposition of a dielectric layer on top of the metallic layer. We will analyze in detail the performance of the magnetoplasmonic interferometers in the presence of this overlayer.

Let us consider again the expression for the modulation of k_x in a Au/Co/Au multilayer system, shown in Eq. 2.20 and reproduced in Eq. 4.1. There it can be seen that the SPP wavevector modulation is proportional to the square of the permittivity of the dielectric layer on top of the metallic multilayer. Therefore, placing dielectric layers with higher ε_d constitutes a simple means of increasing Δk_x. Covering the metallic multilayer with an infinitely thick dielectric other than air is not experimentally feasible, but we can analyze the effect of adding thin dielectric overlayers, which could be seen as adding an effective medium with an intermediate dielectric constant [17]. To verify this, we have covered our Au/Co/Au magnetoplasmonic micro-interferometers by spin coating them with a 60 nm film of polymethyl methacrylate (PMMA) (n = 1.49). Figure 4.6 shows a sketch of the system geometry.

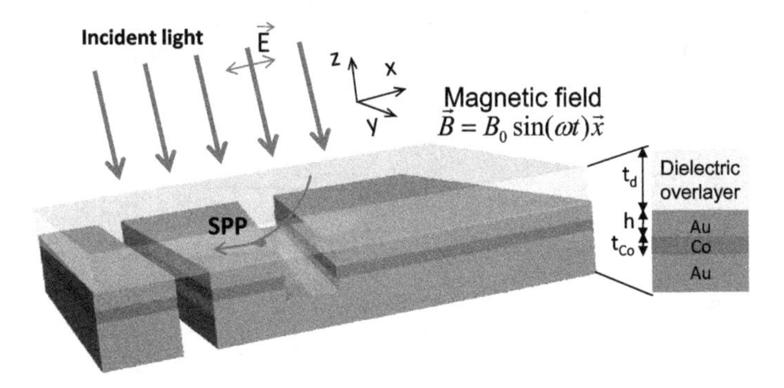

Fig. 4.6 Sketch of the magnetoplasmonic interferometer consisting of a metallic Au/Co/Au trilayer covered by a thin dielectric film

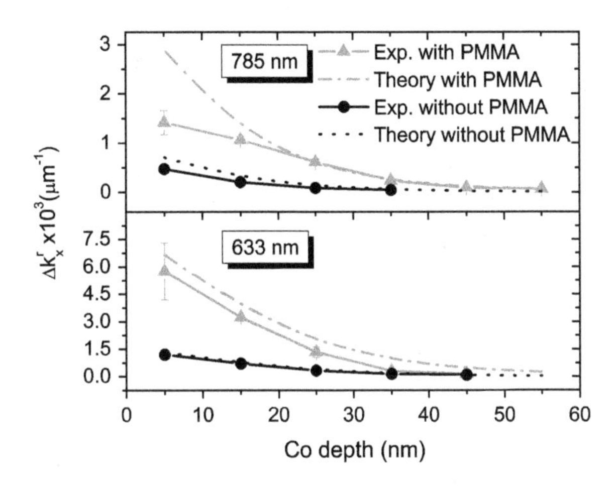

Fig. 4.7 Evolution of the real part of the modulation of the SPP wavevector with the Co layer position for a set of metallic multilayers covered with 60 nm of PMMA and without coverage. Both the experimental data and the simulated curves are plotted. The *upper* (*lower*) *panel* corresponds to a wavelength of 785 (633) nm

Figure 4.7 compares Δk_x for Au/Co/Au trilayers with different h coated by a 60 nm layer of PMMA (triangles) with data without coating (dots) at two different wavelengths, 633 and 785 nm. For those wavelengths, as we have seen in the previous section, we can consider only the real part of the magnetic modulation of the SPP wavevector (Δk_x). We have compared experimental data with theoretical simulations for the two wavelengths. The obtained results for the metallic layer covered by 60 nm of PMMA and without coverage are also plotted in Fig. 4.7, and show an excellent agreement with the experimental values. The SPP wavevector modulation decays exponentially with h, discussed at the beginning of Sect. 4.1. More interestingly, for every Co layer position, Δk_x is higher for the trilayers covered by PMMA, corroborating the theoretical prediction about the SPP modulation enhancement caused by the dielectric layer.

The increase of the magnetoplasmonic SPP modulation due to the addition of a dielectric overlayer can be quantified in terms of the enhancement factor $\Delta k_x^d / \Delta k_x^0$, with Δk_x^d the SPP wavevector modulation for the system covered with a dielectric

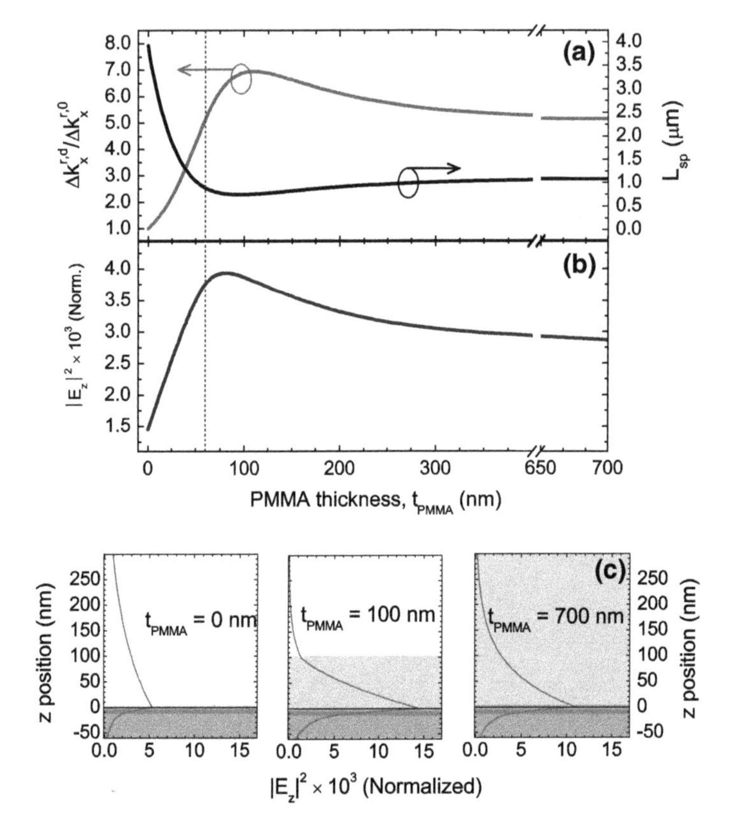

Fig. 4.8 **a** Calculated modulation enhancement (*left axis*) and SPP propagation distance (*right axis*) as a function of the thickness of PMMA coverage for a Au/Co/Au trilayer with $h = 15$ nm. **b** Normalized electromagnetic field intensity calculated at the center of the Co layer. The *dashed line marks* the thickness value used experimentally, 60 nm. **c** Normalized electromagnetic field intensity distribution along the vertical direction for three PMMA thicknesses: 0, 100 and 700 nm. The different background colors account for the different materials in the structure

film and Δk_x^0 the modulation for the uncovered system. From the data shown in Fig. 4.7, we infer an experimental enhancement factor of 4.5 for a 60 nm overlayer thickness when $h = 15$ nm at 633 nm; and of 5 at 785 nm. This enhancement factor will depend on the thickness of the dielectric overlayer. As an illustration, Fig. 4.8a shows the calculated evolution of the modulation enhancement factor as a function of the PMMA thickness for a trilayer with $h = 15$ nm and $\lambda = 633$ nm. Instead of a monotonous increase with the PMMA thickness, Δk_x^d goes through a maximum at around 110 nm, where the enhancement reaches a value of 7, and then decreases to reach a saturation value of 5 around 650 nm. This behavior is again related to the SPP electromagnetic field redistribution, in this case caused by the presence of a dielectric overlayer. The three panels in Fig. 4.8c show that a thin dielectric layer causes a waveguiding effect [17] providing the strongest confinement of SPP electromagnetic field for $t_{\mathrm{PMMA}} \sim 100$ nm. As a consequence, the normalized SPP magnetic field intensity at the position of the cobalt layer (Fig. 4.8b) exhibits a non-monotonous behaviour similar to that of the enhancement factor, supporting our explanation. These two curves are not exactly equivalent because Δk_x also depends on ε_d (Eq. 2.20), which effectively increases as the thickness of the overlayer grows.

4.2.1 Figure of Merit with a PMMA Overlayer. Comparison to Other Modulation Methods

The addition of a dielectric overlayer decreases the propagation distance of SPPs, L_{sp}[17]. Figure 4.8b shows this effect in our system, where we observe that the reduction on L_{sp} is indeed quite strong. This could prevent the application of these dielectric covered Au/Co/Au multilayers in actual devices, so a compromise between the modulation enhancement and the propagation distance of the SPP has to be achieved. We analyze this compromise in terms of the figure of merit $2\Delta k_x \times L_{sp}$. In Fig. 4.9 we plot this product as a function of the dielectric film thickness for metallic trilayers covered by dielectric layers of different refractive indexes. The decrease in L_{sp} is compensated by a much stronger rise in Δk_x resulting into the overall increase for the FOM as a function of dielectric overlayer thickness t_d. Moreover, this increase is higher for materials with a higher refractive index. Taking the values of Fig. 4.5 into consideration, a magnetoplasmonic modulator consisting of a Au/Fe/Au interferometer with a separation distance of $3L_{sp}$ and covered with a dielectric with $n_d = 1.49$ could provide intensity modulations of around 12 % in the optimal spectral range (950 nm). This value is not far from other integrated plasmonic modulator performances reported in the literature based on electro-optical effects such as in Refs. [9, 18], where they show a 15 % of modulation. The first reference has the advantage to be already implemented in SPP waveguide configuration, while the interferometer geometry of the second (and of our case) is still an extended one. However, the switching time response in both cases is quite high compared to the

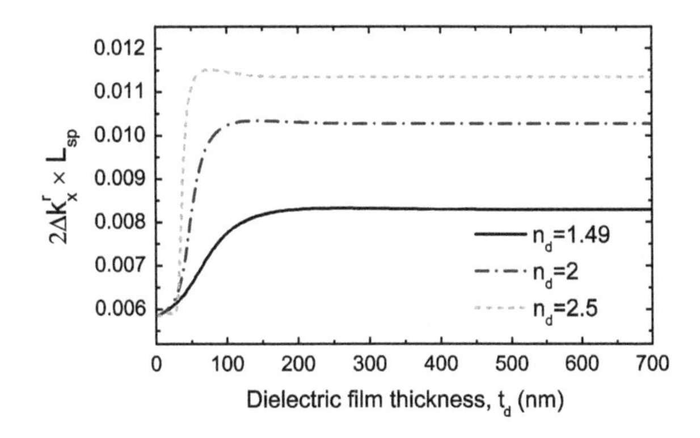

Fig. 4.9 Figure of merit $2\Delta k_x^r \times L_{sp}$ for a Au/Co/Au trilayer with $h = 15$ nm and three different dielectrics

magnetization intrinsic response times (seconds and microseconds for the electro-optical examples compared to femtoseconds for magnetism).

In fact, in the field of active plasmonics, high expectations are placed on electro-optical modulation, since there is a theoretical analysis that predicts modulation depths of 50 % allowing time responses of about 0.01 ns, with very small applied voltages [19], and compatible with integrated CMOS technology. Indeed, some kind of plasmonic transistor has been demonstrated in Ref. [5], with modulation ratios ranging from 16 to 50 % depending on the source-drain distance. Nevertheless, the experimental switching time is 10 μs, and the spectral range of application, very limited to 1.55 μm. Other of the most extended methods to actively modulate plasmons is by thermo-optical control. Some of the first configurations based on it used a Krestchmann configuration and achieved modulation ratios of intensity of about 44 %, although the system was rather bulky [20, 21]. Thermo-optical control was also used in interferometers based on long-range plasmons, being its modulation very large, but the size of the devices was very big (length of the interferometers of several mm) [22]. Later on, those thermo-optical interferometers were implemented with dielectric loaded guides, and modulation depths of about a 20–30 % can be achieved in already integrated devices [10]; however, the main drawback of thermo-optical control is the speed of the modulation, which is of milliseconds or even larger. Smaller although reasonable modulation ratios ($<$10 % or even larger) can also be obtained by using tunable waveguides based on liquid crystals [23, 24], but again the speed is of tens of μs, and they are very large devices. An specially interesting active device is that consisting of ultrafast optical control of SPP, where the modulation speed is in the femtosecond timescale, and modulation depths of about 7.5–35 % have been experimentally seen [25, 26]. It works in an aluminum/silica interface, but its spectral application range is very limited, being besides the spectral dependence of the modulated intensity too dramatic ($<$10 % for 765 nm and $>$30 % for 810 nm). Nowadays, active control of SPPs is being studied in graphene [27–29], since graphene also sustains surface plasmons [29] and the tunability of the properties of graphene sheets makes it a promising material. Indeed, some experiments have been done, and extinction rates of about 6 % have been obtained using a Si grating.

As it can be seen, each method has its advantages and its disadvantages, and depending on the desired application, we could choose one or another. Our magnetoplasmonic interferometer provides small modulation depths compared to other methods; however, it is one of the fastest, it is very easy to implement (there is no need of using special equipment), and it can be integrated. Moreover, in most of the previously mentioned examples (and in the literature), the modulation is in the dielectric material, which makes them suitable for circuitry applications, but inadequate for sensing, where usually the sensing layer is precisely the dielectric. This is not the case for our structures, since the active material is located at the metallic layer (although this is not a required condition in magnetoplasmonics). Besides, and very important, magnetoplasmonic effects are non-reciprocal, which means that the magnetic effect depends on the direction of the traveling plasmon, which could lead to applications such as isolators [30, 31], very relevant in the design of integrated devices.

4.3 Conclusions

Within this chapter we have determined both the real and the imaginary parts of the magnetic field induced modification of the SPP wavevector for Au/FerromagneticMetal/Au magnetoplasmonic systems. It has been shown that the real part is the dominating component in most of the useful spectral range, although at longer wavelengths the values of both components approach and the imaginary part has to be taken into account to appropriately describe the system response. The spectral dependence of the SPP wavevector modulation has been characterized, and it shows a decreasing behavior with wavelength. By means of an analytical expression for Δk_x obtained in the approximation of a very thin ferromagnetic metal layer, we have established that this spectral trend is due to the evolution of the SPP properties with wavelength. These properties come from the vertical confinement of the SPP field, which can be qualitatively described by the evolution of the separation between the SPP wavevector and the light line.

The figure of merit combining both magnetic modulation and propagation distance of the SPP is also analyzed. In terms of spectral dependence, the decrease of SPP wavevector modulation is overcompensated by the increase in L_{sp} for a significant wavelength range, so that the 700 nm–1 μm interval becomes the optimal one for applications. We have extended this analysis to other ferromagnetic metals such as Fe or Ni, resulting that the parameters that govern the spectral dependence are still those related to the SPP properties. The figure of merit, on the other hand, is larger when using Fe instead of Co, although we would need larger magnetic fields to saturate the sample.

Moreover, and without any kind of optimization, we have demonstrated that the deposition of a dielectric overlayer on top of noble/ferromagnetic metal multilayers leads to a significant enhancement of the magnetic field induced modulation of the SPP wavevector. The analysis of the FOM shows that the modulation depth of a magneto-plasmonic switch can be increased despite of the strong reduction of SPP propagation length, which allows to reduce the size of the device. Therefore this finding represents an essential step towards miniaturization of active magneto-plasmonic devices.

Considering all the explained above, a magnetoplasmonic modulator consisting of a Au/Fe/Au interferometer with a separation distance of $3L_{sp}$ and covered with a dielectric with $n_d = 1.49$ could provide intensity modulations of around 12 % in the optimal spectral range (950 nm).

References

1. V.V. Temnov, G. Armelles, U. Woggon, D. Guzatov, A. Cebollada, A. Garcia-Martin, J.M. Garcia-Martin, T. Thomay, A. Leitenstorfer, R. Bratschitsch, Nat. Photonics **4**, 107–111 (2010)
2. D. Martin-Becerra, V.V. Temnov, T. Thomay, A. Leitenstorfer, R. Bratschitsch, G. Armelles, A. Garcia-Martin, M.U. Gonzalez, Phys. Rev. B **86**, 035118 (2012)
3. S.A. Maier, *Plasmonics: Fundamentals and Applications* (Springer, Berlin, 2007)
4. A.V. Krasavin, K.F. MacDonald, N.I. Zheludev, A.V. Zayats, Appl. Phys. Lett. **85**, 3369–3371 (2004)
5. D. Pacifici, H.J. Lezec, H.A. Atwater, Nat. Photonics **1**, 402–406 (2007)
6. R.A. Pala, K.T. Shimizu, N.A. Melosh, M.L. Brongersma, Nano Lett. **8**, 1506–1510 (2008)
7. J.A. Dionne, K. Diest, L.A. Sweatlock, H.A. Atwater, Nano Lett. **9**, 897–902 (2009)
8. A. Agrawal, C. Susut, G. Stafford, U. Bertocci, B. McMorran, H.J. Lezec, A.A. Talin, Nano Lett. **11**, 2774–2778 (2011)
9. M.J. Dicken, L.A. Sweatlock, D. Pacifici, H.J. Lezec, K. Bhattacharya, H.A. Atwater, Nano Lett. **8**, 4048–4052 (2008)
10. J. Gosciniak, S.I. Bozhevolnyi, T.B. Andersen, V.S. Volkov, J. Kjelstrup-Hansen, L. Markey, A. Dereux, Opt. Express **18**, 1207–1216 (2010)
11. J.F. Torrado, J.B. Gonzalez-Diaz, A. Garcia-Martin, G. Armelles, New J. Phys. **15**, 075025 (2013)
12. P.B. Johnson, R.W. Christy, Phys. Rev. B **6**, 4370–4379 (1972)
13. P.B. Johnson, R.W. Christy, Phys. Rev. B **9**, 5056–5070 (1974)
14. G.S. Krinchik, J. Appl. Phys. **35**, 1089–1092 (1964)
15. E.D. Palik, *Handbook of Opical Constants of Solids* (Academic Press, San Diego, 1998)
16. K. Mok, G.J. Kovács, J. McCord, L. Li, M. Helm, H. Schmidt, Phys. Rev. B **84**, 094413 (2011)
17. T. Holmgaard, S.I. Bozhevolnyi, Phys. Rev. B **75**, 245405 (2007)
18. S. Randhawa, S. Lacheze, J. Renger, A. Bouhelier, R.E. de Lamaestre, A. Dereux, R. Quidant, Opt. Express **20**, 2354–2362 (2012)
19. W. Cai, J.S. White, M.L. Brongersma, Nano Lett. **9**, 4403–4411 (2009)
20. A.L. Lereu, A. Passian, J.-P. Goudonnet, T. Thundat, T.L. Ferrell, Appl. Phys. Lett. **86**, 154101 (2005)
21. A. Passian, A.L. Lereu, R.H. Ritchie, F. Meriaudeau, T. Thundat, T.L. Ferrell, Thin Solid Films **497**, 315–320 (2006)
22. T. Nikolajsen, K. Leosson, S.I. Bozhevolnyi, Appl. Phys. Lett. **85**, 5833–5835 (2004)
23. W. Dickson, G.A. Wurtz, P.R. Evans, R.J. Pollard, A.V. Zayats, Nano Lett. **8**(1), 281–286 (2008)
24. D.C. Zografopoulos, R. Beccherelli, Plasmonics **8**, 599–604 (2013)
25. K.F. MacDonald, Z.L. Samson, M.I. Stockman, N.I. Zheludev, Nat. Photonics **3**, 55–58 (2009)
26. Z.L. Samson, K.F. MacDonald, N.I. Zheludev, J. Opt. A Pure Appl. Opt. **11**, 114031 (2009)
27. V.W. Brar, M.S. Jang, M. Sherrott, J.J. Lopez, H.A. Atwater, Nano Lett. **13**, 2541–2547 (2013)
28. W. Gao, G. Shi, Z. Jin, J. Shu, Q. Zhang, R. Vajtai, P.M. Ajayan, J. Kono, Q. Xu, Nano Lett. **13**, 3698–3702 (2013)
29. J. Chen, M. Badioli, P. Alonso-Gonzalez, S. Thongrattanasiri, F. Huth, J. Osmond, M. Spasenovic, A. Centeno, A. Pesquera, P. Godignon, A. Zurutuza-Elorza, N. Camara, F.J. Garcia de Abajo, R. Hillenbrand, F.H.L. Koppens, Nature **487**, 77–81 (2012)
30. J.B. Khurgin, Appl. Phys. Lett. **89**, 251115 (2006)
31. Z. Yu, G. Veronis, Z. Wang, S. Fan, Phys. Rev. Lett. **100**, 023902 (2008)

Chapter 5
Analysis of the Sensing Capability of Plasmonic and Magnetoplasmonic Interferometers

In this chapter, the use of our interferometers as biological sensors is proposed and analyzed. A theoretical comparison of the sensitivity of both plasmonic and magnetoplasmonic interferometers and SPR sensors is shown.

As it has been mentioned in Chap. 2, surface plasmon resonances are commonly used for sensing due to their high sensitivity to changes occurring at the interface in which they take place. Although in the last years there have been several advancements in the use of metallic nanoparticles supporting localized surface plasmons for sensing applications [1–3], traditionally the term Surface Plasmon Resonance (SPR) sensors denotes those based on thin films with propagating surface plasmon polaritons. In fact, SPR sensors are one of the most popular sensing methods, commercialized by several companies, and they are applied in many areas [4, 5], being label-free biosensing one of the most attractive ones [6]. Nowadays, the main goals in the development of SPR sensors lie within the improvement of the sensitivity and the limit of detection [7], as well as the miniaturization [8]. Within this context, some variations of the standard SPR technique have been developed, introducing for example magnetic field (MOSPR) [9] or photonic waveguides [10].

Interferometry in itself has important applications, being for this chapter sensing and biosensing [11, 12] the interesting ones. Interferometry constitutes an important route for developing compact integrated sensors, such as Mach–Zehnder configurations in silicon [13] or in polymer [14]. In fact, interferometry based sensors have already been compared to SPR techniques, and the former ones have demonstrated higher sensitivity in most of the situations [11], although SPR sensors are still highly competitive taking into account the ease of use and the extended knowledge of immobilization protocols in gold. We cited in Chap. 3 the existence of several works on plasmonic interferometers [15–17]. Indeed, plasmonic interferometry for sensing has also been demonstrated, both theoretically and experimentally, in different configurations [18–24]. Although these works show that the plasmonic interferometers offer a good performance as sensors, a direct comparison with the traditional SPR configuration has not been carried out yet.

On the other hand, modulation techniques are usually applied to increase the signal-to-noise ratio in small signals and can be employed to increase the limit of

D. Martín Becerra, *Active Plasmonic Devices*, Springer Theses,
DOI 10.1007/978-3-319-48411-2_5

detection and the sensitivity for different sensing systems. In particular, for SPR systems, mechanical [25], phase [26] or magnetic field modulations [9, 27] have been implemented, demonstrating an increase in sensitivity compared to non-modulated configurations. Plasmonic interferometers, as we know, also allow the introduction of modulated configurations, such as magnetooptical [28–30], all-optical [16] or electrooptical [31] ones. For all the all-optical and electrooptical cases, these modulations are originated at the dielectric material, and are therefore not suitable for sensing applications since the dielectric constitutes the sensing material or analyte. In magnetically modulated plasmonic interferometers, on the other hand, the active material is a ferromagnetic metal, which avoids the aforementioned problem. Moreover, the magnetic modulation of the SPP presents a quite large dependence on the refractive index of the dielectric material [29], as it has been shown on the last part of Chap. 4, which indicates that MP interferometers are a promising tool for sensor development.

Following the path inspired by these results, in this chapter we propose the use of the previously explained MP interferometers as sensing devices, and we theoretically analyze their sensitivity compared to plasmonic (non-magnetic) interferometers. Moreover we carry out first a detailed comparison of these last ones with the traditional SPR technique.

5.1 General Description of Sensing Layers and Methodology

Within this chapter we are going to compare, as it has been said above, three sensing techniques: the two interferometries explained in Chap. 3 (Plasmonic and MP interferometry) and SPR. In this case, all the analysis has been carried out on a numerical basis, using the transfer matrix formalism described in Sect. 3.5 that includes magneto-optical effects [32, 33]. We will start by establishing the needed definitions and background to compare the performance of the three sensors.

Figure 5.1 shows schematically the three compared sensing techniques. Let's see briefly their similarities and differences. SPR (Fig. 5.1a) is based on exciting a surface plasmon in a thin metallic layer (usually ∼50 nm Au layer) by means of the attenuated total internal reflection (ATR) configuration described in Sect. 2.1.1 and measuring the corresponding minimum in the reflectivity R. Both interferometries (Fig. 5.1b, c), on the other hand, require a thick metallic layer (usually ∼200 nm thick) and a defect such as a groove to launch the SPP. As it is explained in Chap. 3, they are based on the existence of interferences between the SPP and the light directly transmitted through a slit cutting the metal film. In a plasmonic interferometer the metallic film is a noble metal (such as gold) and we measure directly the intensity of these interferences I_o, while for MP interferometry a thin ferromagnetic film (such as Co) is inserted to allow magnetic modulation, and we measure the magnetically modulated interferences intensity I_{mp}.

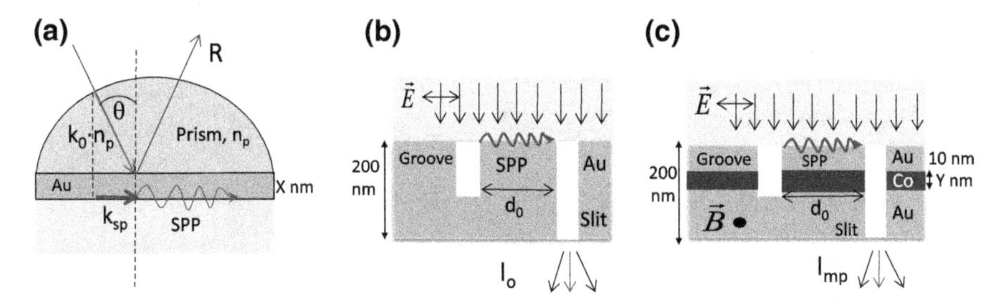

Fig. 5.1 **a** Schematic representation of the traditional SPR configuration. The thickness of the gold layer, X, ranges from 32 to 50 nm depending on the wavelength, **b** sketch of the plasmonic interferometer, **c** schema of the MP interferometer. The thickness of the Co layer, Y, depends on the wavelength, varying from 5 to 10 nm

Table 5.1 Optimized thicknesses for the metallic layers of the SPR and the MP interferometer sensors

Wavelength	X Au (nm)	Y Co (nm)
	SPR	MP interferometer
550	32	5
600	42	5
633	46	6
650	47	6
750	50	6
850	49	8
950	47	10

5.1.1 *Metal Thickness Optimization*

As it can be seen in Fig. 5.1 and has been briefly explained above, the three sensors use different metallic layers. To perform a fair comparison, we have theoretically optimized the thickness of the metallic layers for each configuration and wavelength using the same noble metal, Au, in the three cases. The optimized parameters appear in Table 5.1. For conventional SPR, we have chosen SF10 glass ($n_{SF10} = 1.73$) as substrate and we have calculated the Au layer thickness that provides the optimum SPP excitation, which is that providing the reflectivity closer to zero under ATR configuration at the SPP angle excitation [34]. The obtained Au thickness values range from 32 to 50 nm. In the case of plasmonic interferometry, we need an optically opaque gold layer, so an appropriate thickness is 200 nm Au layer [35] independently of the wavelength. Finally for the magnetoplasmonic MP interferometer, we look for the maximum figure of merit (defined in Sect. 3.6), which is controlled by the thickness and position of the Co layer. The optimum position of the Co layer is as close to the surface as possible, but in an experimental implementation this could cause oxidation so we have fixed it at 10 nm, which is a reasonable value to prevent oxidation in sensing aqueous environment. The complete metallic multilayer for the

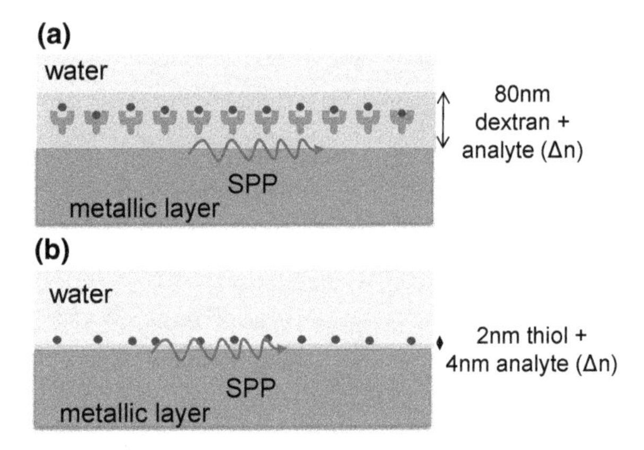

Fig. 5.2 Simulated sensing configurations. **a** "Bulk" system: metallic layer covered by a 80 nm dextran layer (represented by antigen-antibody pairs) immersed in water solution, **b** "surface" system: metallic layer covered by a 2 nm thiol layer and a 4 nm analyte layer (represented by small molecules) immersed in water solution

MP interferometers is then 10Au/*Y*Co/180Au, where the thickness of the cobalt layer varies from 5 to 10 nm.

5.1.2 Sensing Layer

Sensing can be performed in different environments depending on the analyte. One of the most commonly used ones, mainly in biosensing, is an aqueous solution. Although each analyte and sensing protocol requires a specific geometry of the sensing layer, in our theoretical study we have chosen two generic situations that reproduce two of the most common configurations: two-dimensional ("surface") immobilization of biorecognition elements [7] and absorption of the analyte in a three-dimensional matrix ("bulk") [7, 36]. Figure 5.2a represents the theoretically simulated "bulk" configuration that consists of a 80 nm layer of hydrated carboxymethyl dextran, an hydrogel used for biosensing with gold in SPR and whose refractive index in aqueous solution is 1.36 [5, 37]. A sensing experiment will be then simulated as a uniform 80 nm medium whose refractive index (n) is varied from its original refractive index $n_0 = 1.36$ up to 1.363 ($\Delta n = 3 \times 10^{-3}$), mimicking the changes induced by the presence of an attached analyte. On the other hand, the "surface" configuration, shown in Fig. 5.2b, consists of a 2 nm self-assembled monolayer of thiol (a standard ligand with $n = 1.5$) with a uniform 4 nm overlayer whose refractive index will vary from 1.33 (water) up to 1.333, mimicking the adhesion of molecules such as streptavidin to the thiol layer.

5.1.3 Methodology

Finally, the three analyzed sensing techniques use different sensor outputs, which makes their comparison subtle. We will focus on the sensitivity of the sensor, i.e.

the dependence of the measured output, O, with the refractive index of the dielectric medium, n. In a plasmonic system, the sensitivity can be described as [6, 38]:

$$S \equiv \frac{dO}{dn} = \frac{\partial O}{\partial k_x}\frac{\partial k_x}{\partial n} \tag{5.1}$$

In this expression, together with the sensor output and n, it also appears the physical parameter being modified by the change in the refractive index, which in a plasmonic sensor usually corresponds to the SPP wavevector, k_x. In fact, Eq. 5.1 shows that the sensitivity S can be decomposed in two terms: the variation of the actual output measured in the experiment with the physical parameter; and the variation of this physical parameter with the refractive index. The first term is dependent on the measurement method, whereas the other term depends on the used materials and geometry and the properties of the associated SPP. In order to differentiate the effect on the sensitivity ascribed to the measuring technique from those related to the properties of the physical parameter, in our comparison we will analyze separately the second term and the complete dependence of O regarding n for the three sensors. Moreover, as SPP properties depend on the wavelength, we will perform this comparison for different wavelengths.

5.2 SPR Versus Plasmonic Interferometer

5.2.1 Dependence of k_x on n

We will start our analysis comparing the SPR performance with that of the plasmonic interferometer. Both sensors are based on surface plasmon polaritons, defined by its wavevector, k_x. For an interface of two semi-infinite materials, we saw in Sect. 2.1.1 that k_x is expressed as [39]:

$$k_x = k_0 \cdot n_d \sqrt{\frac{\varepsilon_m}{\varepsilon_m + {n_d}^2}}, \tag{5.2}$$

where we have substituted ε_d by the refractive index of the dielectric material, n_d ($\varepsilon_d = n_d^2$). When the refractive index of the dielectric material (our sensing layer) changes from n_0 to n, the physical parameter involved in both techniques is therefore the SPP wavevector, whose modification can be expressed as:

$$\Delta^n k_x = k_x(n) - k_x(n_0) = \frac{\partial k_x}{\partial n} \cdot \Delta n, \tag{5.3}$$

where $\Delta n = n - n_0$.

Equation 5.2 can only be applied to an interface composed of two semi-infinite materials, but given the evanescent nature of the SPP it is a good approximation for

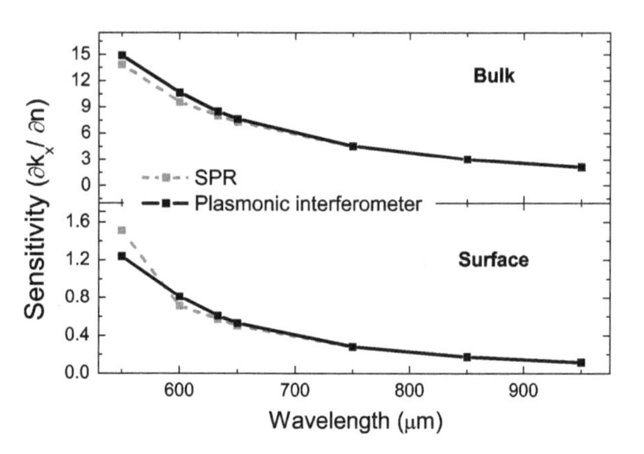

Fig. 5.3 Sensitivity to the refractive index of the SPP wavevector $\left(\frac{\partial k_x}{\partial n}\right)$ for the SPR and the plasmonic interferometer as a function of the wavelength. Both "bulk" (*upper graph*) and "surface" (*lower graph*) configurations are shown

the thick metallic layer of the plasmonic interferometer. However, for the thin layer composing the SPR sensor, the SPP is affected by the presence of the substrate and its wavevector is slightly different (no analytical expression for k_x can be obtained in this case, it has to be calculated numerically). As the wavevectors are different for the thin and thick metal film cases, the sensitivity of k_x for the two sensing systems may also differ. Moreover, both the "surface" and "bulk" configurations have to be considered in order to investigate possible differences associated with the SPP spreading across the sensing layer between the two systems, which would affect their sensitivity differently for the two situations. Besides, being the dielectric constant of metals dispersive, k_x and consequently its sensitivity also depend on wavelength so we also need to take into account this factor.

The analysis of the sensitivity of the SPP wavevector ($\frac{\partial k_x}{\partial n} = \frac{\Delta^n k_x}{\Delta n}$ for small Δn) for SPR and plasmonic interferometry sensors as a function of the wavelength, for both "bulk" and "surface" configurations, is shown in Fig. 5.3. It can be seen that the two techniques present a quite close sensitivity in both "surface" and "bulk" configurations. The "bulk" configuration is more sensitive than the "surface" one, as expected [7], since the spreading of the SPP is much larger than the sensed region for the "surface" system. However, the important result here is that the sensitivity is the same for the two analyzed techniques in each sensing system, meaning that both metallic layers are equally appropriate for sensing bulk changes or surface ones. Regarding the wavelength dependence, both techniques show the same behavior: the SPP wavevector sensitivity increases for lower wavelengths [7, 39]. This is related to the spreading of the SPP electromagnetic field as a function of the wavelength, as it has been explained in Sect. 2.1.1. For shorter wavelengths, the SPP field is more confined to the interface and the system is therefore more sensitive to changes taking place there. Summarizing, the physical parameter involved in both SPR and plasmonic interferometry, k_x shows the same sensitivity in both cases. However the final sensitivity of a sensing technique also depends on how the variation of k_x is translated into a variation of the measured parameter O, as stated by Eq. 5.1. In the following we will thus analyze the sensitivity of O.

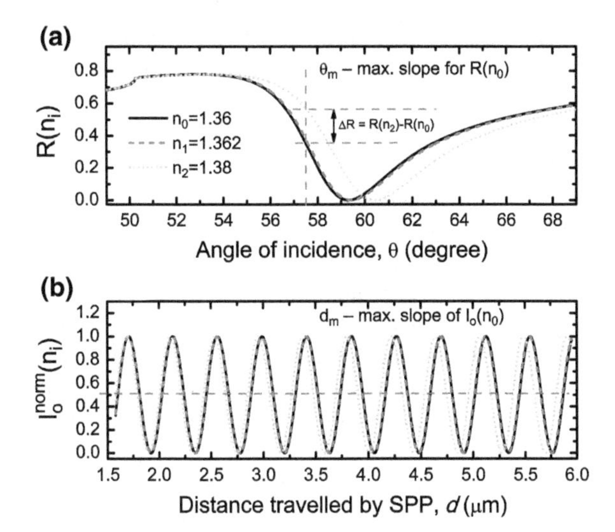

Fig. 5.4 **a** Reflectivity as a function of the angle of incidence for different refractive index n_i in a standard SPR setup, **b** intensity of the plasmonic interferometer, normalized to the amplitude of the oscillations, as a function of the distance d traveled by the plasmon for different refractive index n_i. In the calculation, both I_r (directly transmitted light contribution) and I_{sp} (SPP contribution) are taken to be equal. Both graphs correspond to $\lambda_0 = 633$ nm and to the "bulk" configuration

5.2.2 *Sensitivity of the SPR*

SPR technique, as briefly explained in Sect. 5.1, measures the reflectivity R when exciting the SPP in ATR configuration. To do this, usually a semi-circular prism is set in contact with the metal, as shown in Fig. 5.1a. We saw in Sect. 2.1.1 that for the surface plasmon to be excited, the angle of incidence of the light, θ, must fulfill $k_x = k_0 n_p \sin\,\theta$, being n_p the refractive index of the prism ($n_p = 1.73$ in our simulations). When measuring the reflectivity as a function of θ, a minimum appears when the SPP is excited.

In a sensing experiment, the position of this minimum changes when n varies, because of the associated modification of the plasmon wavevector. In Fig. 5.4a we present a series of reflectivity curves, calculated for different refractive indexes of the sensing layer at a given wavelength $\lambda_0 = 633$ nm. The change in the reflectivity minimum with n can be clearly seen.

A typical procedure in a SPR experiment is to monitor the reflectivity under varying conditions that modify n (e.g. with and without analyte) for a fixed set of wavelength and angle:

$$\Delta R = R(n) - R(n_0). \tag{5.4}$$

We would like to note here that measuring the variation of the reflectivity (which is the method explained here) is not the only procedure for SPR. It can also be done by measuring the shift in the angle of minimum reflectivity, or by analyzing the reflectivity as a function of the wavelength for a given angle and measuring the wavelength shift of the minimum of the reflectivity. However, there are studies that show that the final sensitivity of the SPR technique is independent of the measurement procedure [38], being in all cases affected by the experimental noise in the same manner. Thus, the only differences when comparing the different procedures are due to the precision of the apparatus. As we cannot take into account this in a theoretical treatment, we are going to limit our analysis to the above described reflectivity variation procedure for SPR, without loss of generality. To obtain the maximum sensitivity with this procedure, it is necessary to carefully choose the angle of measurement, θ_m: the maximum ΔR is obtained for the angle that maximizes the slope of the R versus θ curve (see Fig. 5.4a).

To get an insight on the evolution of the reflectivity with n, we can approximate the resonance dip by a Lorentzian curve [38, 40]. This approximation holds quite accurately when $|(\varepsilon_m^r)| \gg \varepsilon_d$ and the metal has low losses, i.e. $\varepsilon_m^i \ll |\varepsilon_m^r|$. The reflectivity around the SPP dip can be written as [40]:

$$R = 1 - \frac{4\gamma_i \gamma_r}{\left(k_0 n_p \sin\theta - k_x\right)^2 + (\gamma_i + \gamma_r)^2}, \tag{5.5}$$

where γ_i is the SPP absorption damping coefficient, that is, the one directly related with the metal absorption, and γ_r is the radiation damping coefficient, that is, the SPP losses due to re-emission of light through the prism at the same angle as the excitation takes place. Both γ_i and γ_r and are responsible of the SPP dip width. The angle of measurement can be obtained after Eq. 5.5 by looking for the angle of maximum slope:

$$\theta_m = \arcsin\left(\frac{k_x \pm \frac{\gamma_i+\gamma_r}{\sqrt{3}}}{k_0 n_p}\right) \tag{5.6}$$

Thus, ΔR ($\equiv (\partial R/\partial n)\Delta n$ for small Δn) at θ_m is given by:

$$\Delta R = \frac{3\sqrt{3}}{2}\frac{\gamma_i \gamma_r}{(\gamma_i + \gamma_r)^3}\Delta^n k_x \tag{5.7}$$

In the case when $\gamma_i = \gamma_r$, it can be seen from Eq. 5.5 that the reflectivity value at the minimum is zero, which is also the condition for optimum SPP excitation [40] that we have employed to define the optimum metal thickness for SPR sensors (see Sect. 5.1.1). Then, ΔR can be rewritten as:

$$\Delta R = \frac{3\sqrt{3}}{16\gamma_i}\Delta^n k_x \tag{5.8}$$

These two last equations, although they are not exact, indicate that the sensitivity of the SPR, $\Delta R/\Delta n$, is related to both the sensitivity of k_x and a term related to the width of the reflectivity dip.

5.2.3 *Sensitivity of the Plasmonic Interferometer*

Although there are several geometries to implement a plasmonic interferometer, all the configurations are intrinsically equivalent. Therefore, in our simulations, we will restrict ourselves to the geometry based on the tilted groove-slit pair (see Fig. 5.1b) described in Chap. 3. As it is explained there, in this interferometer the signal obtained is the intensity oscillations along the position of the slit $I_o(d)$ (Eq. 3.2):

$$I_o = I_r + I_{sp} + 2\sqrt{I_{sp}}\sqrt{I_r}\cos(k_x \cdot d + \varphi_0) \tag{5.9}$$

In Fig. 5.4b it is shown an example of a plasmonic interferogram, where the intensity has been normalized to the amplitude of the oscillations ($\equiv(I_o{}^{MAX} - I_o{}^{min}) = 4\sqrt{I_{sp}}\sqrt{I_r}$). The contrast of this interferogram is maximized when I_r and I_{sp} have the same value at the slit, which can be achieved by tuning the position of the incident light regarding the slit-groove pair and the efficiency of the SPP excitation at the groove [41, 42]. When the refractive index, and therefore the SPP wavevector, changes, the interferogram is shifted, as shown in the graph.

Similarly to the SPR system, both changes in the intensity at a fixed position and shifts in the position of the minima-maxima can be employed as the sensor output O without affecting the intrinsic sensitivity of the system. Here, we have selected the variation with the refractive index of the transmitted intensity I_o at a fixed slit position as the quantity monitored for sensing:

$$\Delta^n I_o = I_o(n) - I_o(n_0) \tag{5.10}$$

Again, in order to attain the highest sensitivity, the optimum position to measure the changes of I_o is that of maximum slope of the I_o versus d curve.

If we neglect the attenuation of the SPP while propagating (that is, if we consider the imaginary part of the plasmon wavevector to be very small), the variation of I_o due to the change of the refractive index in the sensing layer, normalized to the amplitude of the measured interferogram, can be written to first order approximation as:

$$\Delta^n I_o^{\text{norm}} \equiv \frac{\Delta^n I_o}{4\sqrt{I_{sp}}\sqrt{I_r}} \approx -1/2\ \Delta^n k_x\ d\ \sin\left(k_x\ d + \varphi_0\right) \tag{5.11}$$

As it can be seen in Eq. 5.11, the sensitivity of the plasmonic interferometer, $S = \Delta I_o^{\text{norm}}/\Delta n$, is proportional to $\Delta^n k_x \times d$, i.e. to the sensitivity of k_x times the distance traveled by the SPP (see Fig. 5.4b). As this distance can be chosen when fabricating the device, in this sensor we have an external method to increase its sensitivity. The distance traveled by the plasmon can be expressed in terms of the SPP propagation length, L_{sp}. For a plasmonic interferometer, where the Au layer is very thick, the SPP damping is due only to metal absorption, so that $k_x^i = \gamma_i$ [40] and $L_{sp} = 1/(2\gamma_i)$ (see Sect. 2.1.1). By expressing d in units of L_{sp}, $d = f \times L_{sp}$, Eq. 5.11 becomes:

$$\Delta^n I_o^{\text{norm}} \approx \frac{-1}{4\gamma_i} f \, \Delta^n k_x \, \sin\left(\frac{k_x f}{2\gamma_i} + \varphi_0\right). \tag{5.12}$$

5.2.4 Comparison of the Sensitivity for SPR and Plasmonic Interferometry

From Fig. 5.5, we can compare the sensitivity S for both SPR and plasmonic interferometer sensors in the case of the "surface" sensing configuration. For the plasmonic interferometer, different values of d are considered. Plasmonic interferometry proves to be more sensitive than SPR for long enough d, because its sensitivity increases linearly with the distance traveled by the plasmon (Eqs. 5.11–5.12). The plasmonic interferometer sensitivity surpasses that of the SPR sensor for $d \geq 1.5\ L_{sp}$ in all the analyzed spectral range relevant for sensing ($\lambda_0 \geq 600$ nm). In fact, this result can also be obtained by comparing directly Eqs. 5.8 and 5.12 ($3\sqrt{3}/16 < f/4$, which leads to $f > 1.3$, valid at long wavelengths, where the approximations of these equations are valid). Moreover, this slit-groove distance is small enough to keep a good contrast in the interferogram. We estimate that the maximum reasonable value

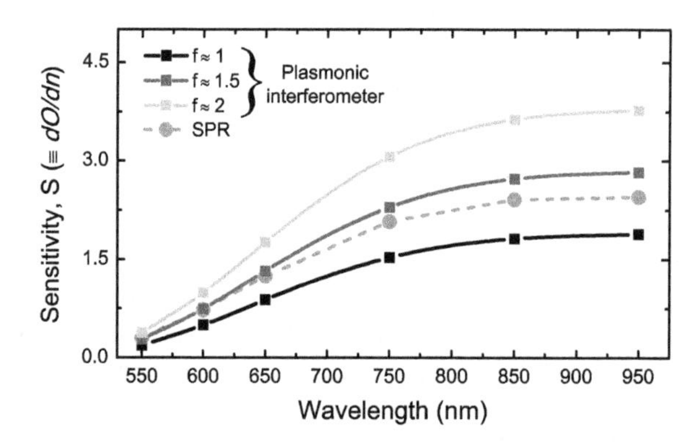

Fig. 5.5 Sensitivity of the SPR ($S = \Delta R/\Delta n$) and the plasmonic interferometer ($S = \Delta^n I_0^{\text{norm}}/\Delta n$) sensors as a function of the wavelength. The calculations have been performed in the "surface" sensing configuration, taking the optimum Au thickness for each wavelength and kind of sensor (200 nm for the plasmonic interferometer and the values collected in Table 5.1 for the SPR). For the plasmonic interferometer, the sensitivity for various slit-groove distances, expressed in terms of $f = d/L_{sp}$, is shown

of d is that providing a minimum interferogram contrast of 0.4 in order to clearly resolve the oscillations. Assuming $I_r = I_{sp}$ at the groove, this results in $d \approx 3\ L_{sp}$. This implies that, by choosing the appropriate slit-groove distance, the performance of plasmonic interferometers for sensing can largely beat that of the conventional SPR. Moreover, Fig. 5.5 also shows that the sensitivity increases with the wavelength for both kinds of sensors. This behavior, opposite to the results presented in Fig. 5.3, reflects the fact that the final sensitivity of a sensor depends not only on the sensitivity of the physical parameter but also on other factors associated with the measured quantity. So, in the case of the SPR sensor, the width of the reflectivity resonance dip also influences S (see Eqs. 5.7–5.8). Since this width decreases strongly with the wavelength, the sensitivity of a SPR system increases at higher wavelengths even though the sensitivity of the physical parameter k_x decreases. Regarding plasmonic interferometry, the SPP propagation length L_{sp} grows rapidly with the wavelength, and being S proportional to the distance traveled by the plasmon (Eqs. 5.11–5.12), this counteracts the tendency of k_x.

5.3 Plasmonic Versus Magnetoplasmonic Interferometry

5.3.1 Evolution of $\Delta^m k_x$ with n

The MP interferometer, described in Chap. 3 and sketched again in Fig. 5.1c, is based on magnetically modulating the plasmon wavevector [28–30]. As explained in detail in Chaps. 2 and 3, when a magnetic field is applied to the MP interferometer in the direction parallel to the surface and perpendicular to the SPP propagation direction, the SPP wavevector is modified as follows [28]:

$$k_x(n, M) = k_x(n, 0) + \Delta^m k_x(n, M) \tag{5.13}$$

where $k_x(n, M)$ and $k_x(n, 0)$ denote the SPP wavevectors in the presence or absence of sample magnetization, respectively, when the dielectric on top of the metal has a refractive index of value n and $\Delta^m k_x(n, M)$ represents the magnetic field induced modification of this SPP wavevector.[1] We have seen in previous chapters that the absolute value of $\Delta^m k_x$ is four orders of magnitude smaller than that of k_x [28–30], but it can be easily measured when using a plasmonic interferometric configuration and inverting synchronously an external magnetic field strong enough to magnetically saturate the Co layer.

When the refractive index changes, both k_x and $\Delta^m k_x$ are modified and therefore the two physical parameters should be taken into account when analyzing the

[1] In this chapter we are going to refer to the magnetical modulation of the SPP wavevector as $\Delta^m k_x$, since we want to clearly distinguish the variations induced by the change in the refractive index (denoted with superindex n) from those induced by the magnetic field (denoted with superindex m).

performance of the MP interferometer as a sensor. The general expression for S in this case becomes:

$$S = \frac{dO}{dn} = \frac{\partial O}{\partial k_x}\frac{\partial k_x}{\partial n} + \frac{\partial O}{\partial \Delta^m k_x}\frac{\partial \Delta^m k_x}{\partial n}. \tag{5.14}$$

Analogously to $\Delta^n k_x$, the variation of $\Delta^m k_x$ when the refractive index of the sensing layer is modified is defined as:

$$\Delta^n\left(\Delta^m k_x\right) = \frac{\partial \Delta^m k_x}{\partial n} \cdot \Delta n \equiv \Delta^m k_x(n, M) - \Delta^m k_x(n_0, M). \tag{5.15}$$

Figure 5.6 shows the evolution of the relative changes of $\Delta^n k_x$ and $\Delta^n(\Delta^m k_x)$ with the wavelength for the plasmonic and the MP interferometers (note that here we need to compare the relative changes of both quantities, and not their absolute values, as these ones differ in several orders of magnitude as mentioned above). The comparison of both quantities offers an interesting result: $\Delta^n(\Delta^m k_x)/\Delta^m k_x$ is one order of magnitude bigger than $\Delta^n k_x/k_x$. This is due to the fact that the magnetic field induced modulation of the SPP wavevector has a strong dependence on the fourth power of the refractive index of the dielectric [29, 30], as it has been shown in Chap. 4. As a consequence, it can be expected that a sensing technique relying on the measurement of the variations of $\Delta^m k_x$ exceeds in sensitivity to another one based on variations of k_x.

From Fig. 5.6 it can also be seen that, for both "surface" and "bulk" sensing configurations, the relative sensitivity of the magnetic parameter surpass in one order

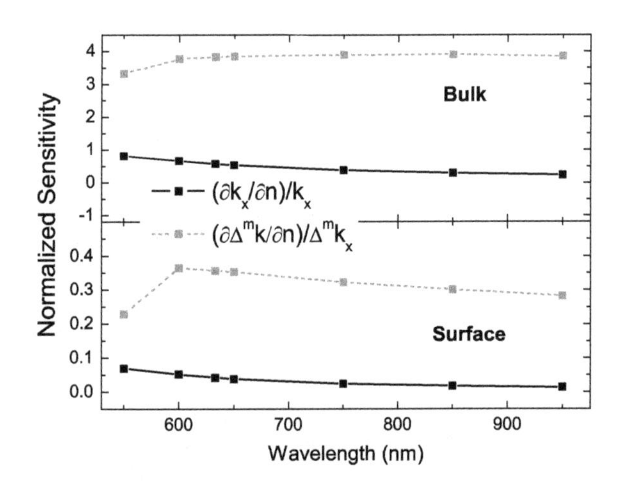

Fig. 5.6 Sensitivity as a function of the wavelength of the normalized SPP wavevector, $\Delta^n k_x/k_x$, for a plasmonic interferometer and of the normalized SPP wavevector magnetic modulation, $\Delta^n(\Delta^m k_x)/\Delta^m k_x$, for a MP one. For each kind of interferometer and wavelength, the optimum metal thickness has been taken. Both "bulk" (*upper graph*) and "surface" (*lower graph*) configurations are shown

of magnitude that of the purely plasmonic one. Regarding the spectral behaviour, except for the smallest wavelength value shown, $\Delta^n(\Delta^m k_x)/\Delta^m k_x$ decreases with the wavelength but in a slower manner than $\Delta^n k_x$, and indeed for the "bulk" configuration it can be considered as nearly constant.

5.3.2 *Sensitivity of the Magnetoplasmonic Interferometer*

From the previous analysis of $\Delta^n(\Delta^m k_x)$, the magnetoplasmonic interferometer seems a promising candidate to develop a sensor. However, care has to be taken in selecting the appropriate measured quantity. As it has been exposed in Chap. 3, a magnetic field applied to the magnetoplasmonic MP interferometer modifies the SPP wavevector, which induces a shift in the transmitted intensity interferogram. This modulation of the intensity, that we have denoted as magnetoplasmonic intensity, I_{mp}, can be expressed, as we have seen, as:

$$I_{mp} = -4\sqrt{I_{sp}}\sqrt{I_r}\ \Delta^m k_x\ d^{mp}\ \sin(k_x\ d^{mp} + \varphi_0), \tag{5.16}$$

if we neglect again the imaginary part of k_x, and considering $\Delta^m k_x\ d^{mp} \ll 1$ [28–30]. We would like to notice here that, for the optimization of the MP interferometer described in Sect. 5.1.1, the quantity we have maximized is I_{mp}. As this quantity is proportional to $\Delta^m k_x \times d^{mp}$, we have indeed maximized our figure of merit, the product $\Delta^m k_x \times L_{sp}^{mp}$[2] (as L_{sp}^{mp} limits the maximum value of d^{mp} that can be used).

One option to use the MP interferometer as a sensor would be the monitorization of I_{mp} at a fixed slit point when the refractive index changes. The variation of this intensity, normalized to its initial amplitude ($\equiv {I_{mp}}^{MAX}(n_0) - {I_{mp}}^{min}(n_0) = 8\sqrt{I_{sp}}\sqrt{I_r}\ \Delta^m k_x(n_0)\ d^{mp}$), can be expressed in first order as:

$$\begin{aligned}
\Delta^n I_{mp}^{\text{norm}} &\equiv \frac{I_{mp}(n) - I_{mp}(n_0)}{8\sqrt{I_{sp}}\sqrt{I_r}\ \Delta^m k_x(n_0)\ d^{mp}} \\
&\approx -1/2\Big[\Delta^n k_x\ d^{mp}\cos\left(k_x\ d^{mp} + \varphi_0\right) \\
&+ \Delta^n\left(\Delta^m k_x\right)/\Delta^m k_x \sin(k_x\ d^{mp} + \varphi_0)\Big] \\
&= -1/2\sqrt{(\Delta^n k_x\ d^{mp})^2 + \left(\Delta^n(\Delta^m k_x)/\Delta^m k_x^2\right)} \\
&\times \sin\left(k_x\ d^{mp} + \varphi_0 + \gamma\right),
\end{aligned} \tag{5.17}$$

[2]It has to be noticed that in our theoretical analysis, the propagation distance of the SPP in the plasmonic system (L_{sp}) is not the same than that of the MP system (denoted therefore as L_{sp}^{mp}), due to the presence of the Co layer. The same is going to happen then for the distance between the slit and the groove (d or d^{mp} respectively).

where $\tan\gamma = \frac{\Delta^n k_x\, d^{mp}\, \Delta^m k_x}{\Delta^n(\Delta^m k_x)}$. Equation 5.17 shows that in fact the MP interferometer is more sensitive than an equivalent plasmonic one. Nevertheless, we should analyze this statement very cautiously. As in the case of the plasmonic interferometer for I_o, the sensitivity of I_{mp} depends on the product $\Delta^n k_x \times d^{mp}$, and therefore the sensitivity can be boosted by increasing the slit-groove distance. This distance is however limited by the SPP propagation distance and, as we discussed above, values of d^{mp} longer than $3L_{sp}^{mp}$ will result in a too reduced contrast for comfortable measurements. If we impose this limit of $d^{mp} \leq 3L_{sp}^{mp}$, the sensitivity of the plasmonic interferometer surpasses that of the MP one as the presence of Co in the metallic trilayer, highly absorbent, strongly reduces the value of L_{sp}^{mp}. This is shown in Fig. 5.7, where the sensitivity of the plasmonic interferometer, relying on the measurement of I_o^{norm}, is compared with that of the MP interferometer based on the monitorization of I_{mp}^{norm}. The sensitivity of a plasmonic interferometer with $d = L_{sp}$ is higher than that of the MP one with $d^{mp} = L_{sp}^{mp}$ or even $d^{mp} = 2L_{sp}^{mp}$ due to the smaller value of L_{sp}^{mp} in this last case (see inset in the figure with the evolution of L_{sp} with the wavelength for the two kind of interferometers). Furthermore, even for a MP interferometer with d^{mp} equal to the value of L_{sp} corresponding to the pure Au layer composing the plasmonic one, which will not be realistic, the sensitivity of I_{mp}^{norm} is comparable or smaller than that of I_o^{norm} (compare light blue dotted line and black solid line in Fig. 5.7). This is due to the fact that the introduction of the Co layer in the MP interferometer also reduces the value of $\Delta^n k_x$, therefore loosing the increase provided

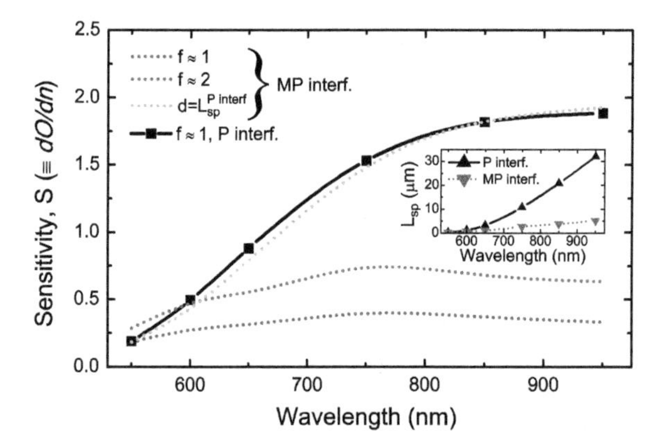

Fig. 5.7 Calculated sensitivity of the measured intensity for the plasmonic ($S = \Delta I_o^{\text{norm}}/\Delta n$) and the MP ($S = \Delta I_{mp}^{\text{norm}}/\Delta n$) interferometers as a function of the wavelength. The calculations have been performed in the "surface" sensing configuration, taking the optimum metal thickness for each wavelength and kind of interferometer (200 nm Au for the plasmonic interferometer and 10Au/*Y*Co/180Au, with the *Y* values collected in Table 5.1 for the MP one). For the magnetoplasmonic interferometer, the sensitivity for various slit-groove distances is shown, expressed in terms of $f = d^{mp}/L_{sp}^{mp}$. The light blue dotted line corresponds to the sensitivity of the MP interferometer for a d^{mp} value equal to the L_{sp} propagation distance of the plasmonic interferometer at the given wavelength. The *inset* shows the L_{sp} (or L_{sp}^{mp}, respectively) values at different wavelengths for the plasmonic and MP interferometers

by the extra term $\Delta^n(\Delta^m k_x)/\Delta^m k_x$ in Eq. 5.17. The alternative of implementing the MP interferometer by monitoring the shifts in the I_{mp} interferogram is not useful in order to take advantage of the higher sensitivity of $\Delta^m k_x$ with respect to k_x either, as the shift is only related to $\Delta^n k_x$ (see Eq. 5.16). A last option that could offer the advantage of higher sensitivity for the MP interferometer is based on the direct monitorization of $\Delta^m k_x$. For that, both the transmitted intensity without a magnetic field applied and the modulated intensity have to be collected for a full oscillation, and by normalizing I_{mp} to the amplitude of the full oscillation of I_o and dividing by the slit-groove distance, the value of $\Delta^m k_x$ is obtained [29, 30]. This full process should be repeated when the refractive index of the sensing layer changes to obtain the new value of $\Delta^m k_x$. This option, although feasible, requires more signal processing and time acquisition, so it could impose some time restrictions in the sensing experiments. However, the higher sensitivity of $\Delta^m k_x$ to the refractive index variations could compensate for this drawback.

5.4 Conclusions

To sum up, the sensitivity of three plasmonic based sensing devices have been compared: standard SPR technique, plasmonic interferometry and MP interferometry. Every system has been considered at its best material implementation, and both the sensitivity of the physical parameter on which the method is based and the actual sensor output have been studied. Table 5.2 provides a summary with all the relevant parameters for each sensor. Both SPR and plasmonic interferometry rely on the modification of the SPP wavevector under refractive index variations. Being the physical system supporting each of them very similar, Au films, there are not significant differences regarding the sensitivity of the SPP wavevector in both cases, neither in a "bulk" sensing configuration or in a "surface" one. When analyzing the measured output, however, it has been shown that the plasmonic interferometer can surpass the SPR in sensitivity. This is due to the fact that the sensitivity of the plasmonic interferometer is proportional to the slit-groove distance, so by increasing this distance the sensitivity can be enhanced. Even though the slit-groove distance is enlarged to improve the plasmonic interferometer sensitivity, it is still kept on the order of a few tens of microns, which allows its integration in a chip for the development of

Table 5.2 Parameters involved in the three different sensing systems analyzed

Sensing method	Physical system	Physical parameter	Measured quantity
SPR	Thin Au film	$\Delta^n k_x$	ΔR
Plasmonic interf.	200 nm Au layer	$\Delta^n k_x$	$\Delta^n I_o$
MP interf.	200 nm	$\Delta^n k_x$ and	$\Delta^n I_{mp}$
	AuCoAu trilayer	$\Delta^n(\Delta^m k_x)$	

miniaturized sensors. Regarding the MP interferometer, two physical parameters are involved in a sensing experiment: the modification of the SPP wavevector and the variation of the magnetic field driven SPP wavevector modulation. This last quantity has a stronger dependence with the refractive index than k_x, so a sensor based on it would provide a higher performance. However, in order to take advantage of this, the monitored output quantity in a MP interferometer has to be directly the processed value of $\Delta^m k_x$.

References

1. J.N. Anker, W.P. Hall, O. Lyandres, N.C. Shah, J. Zhao, R.P. Van Duyne, Nat. Mater. **7**, 442 (2008)
2. M. Svedendahl, S. Chen, A. Dmitriev, M. Kall, Nano Lett. **9**, 4428–4433 (2009)
3. M.A. Otte, B. Sepulveda, W. Ni, J.P. Juste, L.M. Liz-Marzn, L.M. Lechuga, ACS Nano **4**, 349–357 (2010)
4. J. Homola, S.S. Yee, G. Gauglitz, Sens. Actuators B **54**, 3–15 (1999)
5. R.B.M. Schasfoort, A.J. Tudos, *Handbook of Surface Plasmon Resonance* (The Royal Society of Chemistry, Cambridge, UK, 2008)
6. J. Homola, Anal. Bioanal. Chem. **377**, 528–539 (2003)
7. J. Homola, Chem. Rev. **108**, 462–493 (2008)
8. X. Hoa, A. Kirk, M. Tabrizian, Biosens. Bioelectron. **23**, 151–160 (2007)
9. B. Sepulveda, A. Calle, L.M. Lechuga, G. Armelles, Opt. Lett. **31**, 1085–1087 (2006)
10. T. Srivastava, R. Das, R. Jha, Sens. Actuators B **157**, 246–252 (2011)
11. R. Ince, R. Narayanaswamy, Anal. Chim. Acta **569**, 1–20 (2006)
12. A. Kussrow, C.S. Enders, D. Bornhop, J. Anal. Chem. **84**, 779–792 (2012)
13. F. Prieto, B. Seplveda, A. Calle, A. Llobera, C. Domnguez, A. Abad, A. Montoya, L.M. Lechuga, Nanotechnology **14**, 907 (2003)
14. R. Bruck, E. Melnik, P. Muellner, R. Hainberger, M. Lmmerhofer, Biosens. Bioelectron. **26**, 3832–3837 (2011)
15. G. Gay, O. Alloschery, B. Viaris de Lesegno, C. O'Dwyer, J. Weiner, H.J. Lezec, Nature **2**, 262–267 (2006)
16. D. Pacifici, H.J. Lezec, H.A. Atwater, Nat. Photonics **1**, 402–406 (2007)
17. V.V. Temnov, U. Woggon, J. Dintinger, E. Devaux, T.W. Ebbesen, Opt. Lett. **32**, 1235–1237 (2007)
18. X. Wu, J. Zhang, J. Chen, C. Zhao, Q. Gong, Opt. Lett. **34**, 392–394 (2009)
19. Y. Gao, Q. Gan, Z. Xin, X. Cheng, F.J. Bartoli, ACS Nano **5**, 9836–9844 (2011)
20. X. Li, Q. Tan, B. Bai, G. Jin, Opt. Express **19**, 20691–20703 (2011)
21. J. Feng, V.S. Siu, A. Roelke, V. Mehta, S.Y. Rhieu, G. Palmore, R. Tayhas, D. Pacifici, Nano Lett. **12**, 602–609 (2012)
22. O. Yavas, C. Kocabas, Opt. Lett. **37**, 3396–3398 (2012)
23. T. Bian, B.-Z. Dong, Y. Zhang, Plasmonics **8**, 741–744 (2013)
24. Y. Gao, Z. Xin, Q. Gan, X. Cheng, F.J. Bartoli, Opt. Express **21**, 5859–5871 (2013)
25. Y.-B. Shin, H.M. Kim, Y. Jung, B.H. Chung, Sens. Actuators B **150**, 1–6 (2010)
26. P.P. Markowicz, W.C. Law, A. Baev, P.N. Prasad, S. Patskovsky, A. Kabashin, Opt. Express **15**, 1745–1754 (2007)
27. D. Regatos, B. Sepúlveda, D. Fariña, L.G. Carrascosa, L.M. Lechuga, Opt. Express **19**, 8336–8346 (2011)
28. V.V. Temnov, G. Armelles, U. Woggon, D. Guzatov, A. Cebollada, A. Garcia-Martin, J.M. Garcia-Martin, T. Thomay, A. Leitenstorfer, R. Bratschitsch, Nat. Photonics **4**, 107–111 (2010)

29. D. Martin-Becerra, J.B. Gonzalez-Diaz, V. Temnov, A. Cebollada, G. Armelles, T. Thomay, A. Leitenstorfer, R. Bratschitsch, A. Garcia-Martin, M.U. Gonzalez, Appl. Phys. Lett. **97**, 183114 (2010)
30. D. Martin-Becerra, V.V. Temnov, T. Thomay, A. Leitenstorfer, R. Bratschitsch, G. Armelles, A. Garcia-Martin, M.U. Gonzalez, Phys. Rev. B **86**, 035118 (2012)
31. M.J. Dicken, L.A. Sweatlock, D. Pacifici, H.J. Lezec, K. Bhattacharya, H.A. Atwater, Nano Lett. **8**, 4048–4052 (2008)
32. Gonzalez-Diaz, J. B. *MagnetoPlasmonics. MagnetoOptics in Plasmonics Systems*. Ph.D. thesis, (Universidad Autonoma de Madrid, 2010)
33. K. Postava, J. Pistora, S. Visnovsky, Czechoslov. J. Phys. **49**, 1185–1204 (1999)
34. Regatos Gómez, D. *Biosensores Opticos de alta sensibilidad basados en tecnicas de modulacion plasmonica*. Ph.D. thesis (Universidad de Santiago de Compostela, 2012)
35. V.V. Temnov, K. Nelson, G. Armelles, A. Cebollada, T. Thomay, A. Leitenstorfer, R. Bratschitsch, Opt. Express **17**, 8423–8432 (2009)
36. M.G. Manera, G. Montagna, E. Ferreiro-Vila, L. Gonzalez-Garcia, J.R. Sanchez-Valencia, A.R. Gonzalez-Elipe, A. Cebollada, J.M. Garcia-Martin, A. Garcia-Martin, G. Armelles, R. Rella, J. Mater. Chem. **21**, 16049–16056 (2011)
37. J. Piehler, A. Brecht, K. Hehl, G. Gauglitz, Colloids Surf. B. **13**, 325 (1999)
38. M. Piliarik, J. Homola, Opt. Express **17**, 16505–16517 (2009)
39. S.A. Maier, *Plasmonics: Fundamentals and Applications* (Springer, Berlin, 2007)
40. H. Raether, *Surface Plasmons* (Springer, Berlin, 1986)
41. J. Renger, S. Grafström, L.M. Eng, Phys. Rev. B **76**, 045431 (2007)
42. P. Lalanne, J.P. Hugonin, H.T. Liu, B. Wang, Surf. Sci. Rep. **64**, 453–469 (2009)

Chapter 6
Near Field Magnetoplasmonic Interferometry

Here, the concept of magnetically modulating SPP in an interferometric system will be analyzed in the near field regime. Different configurations that allow this will be studied theoretically and a experiment will be proposed.

In previous Chaps. 3 and 4, the magnetic modulation of SPPs has been analyzed in the far field using magnetoplasmonic interferometers [1–3]. Nevertheless, given that SPPs involve evanescent fields, it will be quite interesting studying them (and the effect of the magnetic field) in the near field (NF). We propose in this chapter different configurations to evaluate the magnetic field induced modulation in the near field, paying special attention to the magnitude of the modulation for experimental verification.

The main difference concerning the analysis in the far field that we have seen before is that not only k_x^r and k_x^i, but also k_z [4], the variation in the vertical component of the SPP wavevector, and its magnetic modulation Δk_z, need to be considered in the NF; and thus this magnitude could be obtained from the measurements. First, an interferometric configuration with two counterpropagating plasmons, that is widely used in near field, will be analyzed theoretically. Then, a single SPP will be also analyzed, since a NF technique would theoretically allow us to have access to the modulation of other components of the SPP, not only to k_x^r. Finally, in order to try to have larger signals than for a single SPP, or larger modulation ratios, a more complex interferometric plasmonic system consisting of the interference between a SPP and radiative light is analyzed. The magnetic modulation of the SPP wavevector for all these configurations in the near field will be studied theoretically. Nevertheless the analysis will be done regarding a possible experimental performance; and the estimated values of modulations that could be expected will be shown. Finally, we will show a performed experimental implementation using a scanning near field optical microscopy (SNOM) for one of the interferometric configuration proposed.

D. Martín Becerra, *Active Plasmonic Devices*, Springer Theses,
DOI 10.1007/978-3-319-48411-2_6

6.1 Two Counterpropagating SPPs

The first configuration that we want to discuss is the case of two counterpropagating SPPs, which has often been considered in the near field [5, 6]. An easy way to implement this is to use two parallel slits where the SPP excitation is carried out by illumination of the sample from below [5]. A sketch of the geometry of the system analyzed and the sample can be seen in Fig. 6.1. The sample consists of a 200 nm Au/Co/Au trilayer with 6 nm of cobalt and an upper gold layer of 15 nm (although sometimes we will consider also a Au layer of 10 nm). By covering the two slits with the illumination spot from the glass side, four SPPs, among other waves, will be launched [7, 8] at the Au/air interface, two at each slit. As it can be seen at Fig. 6.1, the two central SPPs, that propagate in opposite directions, will interfere all over the space between the two slits. In this section, we are going to study the effect of a transverse magnetic field (parallel to the surface sample and perpendicular to the SPP propagation direction) into that interference. As it happened in the far field interferometry, the magnetic field will modify the SPP wavevector and therefore the intensity of the interference at each point.

To analyze the near field effect of this modification we will begin with a simplified analysis of the system, in order to define the different quantities involved and to obtain the first estimations of what can be expected. Then, other relevant aspects such as the propagation losses, will be considered. We will first assume that the z component of the SPP electromagnetic field is the dominant one, neglecting the x component, which will be included later. Moreover, two different NF configurations could be considered: the use of tapered dielectric fibers as tips [6], and the use of fluorescent particles [5, 9, 10]. In the first case, the collected intensity corresponds to the square

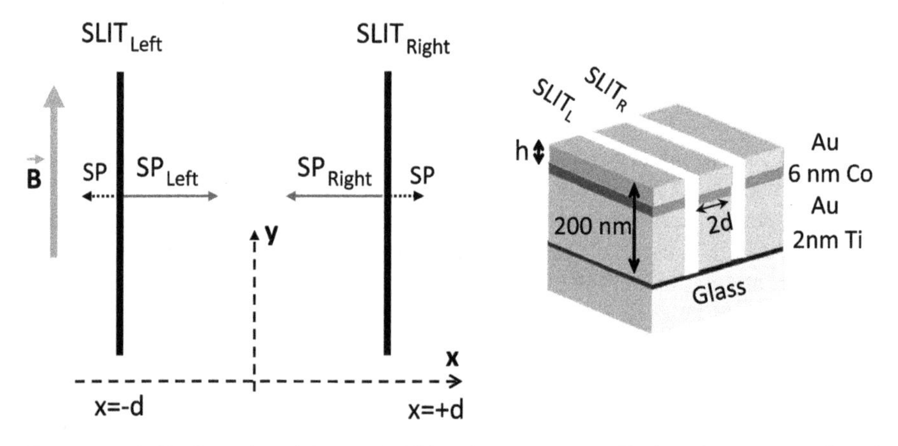

Fig. 6.1 (*Left*) Configuration of the near field interferometer setup. The interferometer consists of two slits crossing all the metallic layer. Between the slits, two counterpropagating SPPs launched at each one of them, SP_{Left} and SP_{Right}, will interfere. The magnetic field will be applied parallel to the slits. (*Right*) The sample consists of a 200 nm Au/Co/Au trilayer with 6 nm of cobalt and an upper gold layer of $h =$ 15 or 10 nm

power of the field, while in the other case it is the fourth power of it. The particle (tip) size has also been taken into account by calculating the integration of the field over its volume. Due to the complexity of the expressions, for several systems we have kept the analytical analysis to the case of field intensity $|E|^2$ and point size tip, to get an idea of the influence of the different parameters, while the E^4 case and the finite size tip have been analyzed only numerically.

6.1.1 *First Approximation, Only One SPP Electromagnetic Field Component: E_z*

In this initial analysis, we are going to consider only the z component of the SPP electromagnetic field, E_z. Besides, we will not take into consideration neither the imaginary part of k_x nor its modulation Δk_x^i, to keep it as simple as possible. The vertical component of the SPP wavevector, k_z, will be considered as purely imaginary $k_z = ik_z^i$, and its modulation will also be discarded. Later on we will analyze whether these are reasonable simplifications or not in the near field. Thus, the E_z component of the electromagnetic field of each plasmon propagating between the two slits can be expressed as:

$$\begin{aligned} E_{z,L}(SP_{Left}) &= A \cdot e^{ik_x^r(x+d)}e^{-k_z^i z}, \\ E_{z,R}(SP_{Right}) &= A' \cdot e^{-ik_x^r(x-d)}e^{-k_z^i z} \end{aligned} \tag{6.1}$$

The subindex L refers to the SPP launched at the left slit propagating towards the right, and R to the one launched at the slit of the right and propagating towards the left. A and A' are the SPP excitation efficiency at each slit. In principle $A \neq A'$ but if the two slits are identical and the illumination is homogeneous regarding the slits, we could consider $A = A'$. These two plasmons interfere between the slits [5], and the interference intensity for the two counterpropagating surface plasmons is given by:

$$I = |E_{z,L} + E_{z,R}|^2 = e^{-2k_z^i z}\left[A^2 + A'^2 + 2AA'\cos(2k_x^r \cdot x)\right] \tag{6.2}$$

An example of this intensity pattern can be seen in Fig. 6.2. Here, the light wavelength is $\lambda_0 = 980\,\text{nm}$; the vertical distance from the metal surface is $z = 400\,\text{nm}$; the distance between slits, $2d$, is $15\,\mu\text{m}$; $h = 15\,\text{nm}$, and $A = A' = 0.5$.

The next step consists of introducing the magnetic field parallel to the slits. The dependence of the SPP wavevector with the applied magnetic field is given by Eq. 2.14: $k_x(\vec{M}) = k_x^0 + \Delta k_x \cdot m$. Keeping in mind that the sign of the effect of the magnetic field depends on the relative direction between $\vec{k_x}$ and $\vec{M}$, the electromagnetic field of each plasmon becomes:

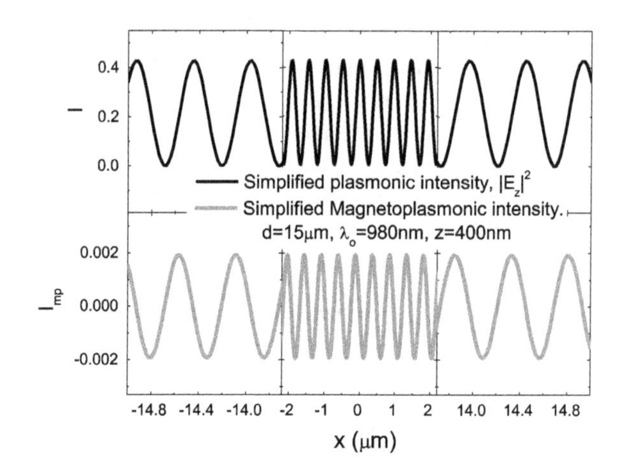

Fig. 6.2 Plasmonic intensity and MP intensity for two counterpropagating SPP, being $d = 15\,\mu\text{m}$, $h = 15\,\text{nm}$, and $\lambda_0 = 980\,\text{nm}$ and $z = 400\,\text{nm}$. We have considered $A = A' = 0.5$. Be aware of the different scales for x axis

$$\begin{aligned}
&\text{For } \vec{M}^+ = +M \cdot \vec{y}: \\
E_{z,L}(+M) &= e^{-k_z^i z} A \cdot e^{i(k_x^r + \Delta k_x^r)\cdot(x+d)}, \\
E_{z,R}(+M) &= e^{-k_z^i z} A' \cdot e^{-i(k_x^r - \Delta k_x^r)(x-d)} \\
&\text{For } \vec{M}^- = -M \cdot \vec{y}: \\
E_{z,L}(-M) &= e^{-k_z^i z} A \cdot e^{i(k_x^r - \Delta k_x^r)\cdot(x+d)}, \\
E_{z,R}(-M) &= e^{-k_z^i z} A' \cdot e^{-i(k_x^r + \Delta k_x^r)(x-d)}
\end{aligned} \tag{6.3}$$

The interference intensity then reads as:

$$\begin{aligned}
I(+M) &= e^{-2k_z^i z}\left[A^2 + A'^2 + 2AA'\cos\left(2k_x^r \cdot x + 2\Delta k_x^r \cdot d\right)\right] \\
I(-M) &= e^{-2k_z^i z}\left[A^2 + A'^2 + 2AA'\cos\left(2k_x^r \cdot x - 2\Delta k_x^r \cdot d\right)\right]
\end{aligned} \tag{6.4}$$

In analogy to the far field treatment, we define the MP intensity in the near field as $I_{mp} = I(+M) - I(-M)$. Thus, considering $\Delta k_x^r d \ll 1$, it can be expressed as:

$$I_{mp} \approx -8AA'\Delta k_x^r \cdot d e^{-2k_z^i z} \sin\left(2k_x^r \cdot x\right) \tag{6.5}$$

As it can be seen from this equation, I_{mp} is a sinusoidal pattern phase shifted by 90° regarding the plasmonic wave pattern I (see Fig. 6.2). Notice that exactly in the middle of the two slits, there is no modulation of the intensity. This is due to the fact that the system is completely symmetric, therefore the modulation of the SPP traveling from the left compensates with the SPP traveling from the right, since they have covered the same distance but in opposite directions.

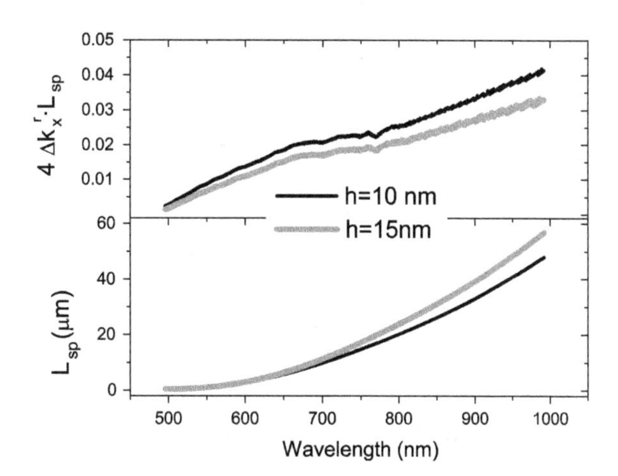

Fig. 6.3 Figure of merit for two counterpropagating SPPs in the near field as a function of the wavelength for a Au/Co/Au trilayer with $h = 10$ and $h = 15\,\text{nm}$

In the far field experiments, the measurements involved obtaining both I_{mp} and I simultaneously. After that, we calculated the ratio between both contrasts in order to finally obtain the magnetic modulation of the SPP wavevector, Δk_x^r (as in Chap. 3). The same treatment can be applied here. If we obtain the ratio of the contrasts of I_{mp} and I we get:

$$\frac{\text{contrast}(I_{mp})}{\text{contrast}(I)} = 4\Delta k_x^r \cdot d \tag{6.6}$$

In fact, for $\lambda = 980\,\text{nm}$, a distance between slits $2d = 15\,\mu\text{m}$ and with an upper Au layer of 15 nm this ratio is about $9 \cdot 10^{-3}$, as it can be seen from Fig. 6.2. In analogy to the far field case, the modulated magnitude here is the NF intensity and in this case Δk_x^r is not the only important parameter, given that it always appears together with the separation between slits, $2d$. Therefore, again here we can consider a figure of merit, defined as $4\Delta k_x^r L_{sp}$, to analyze the expected modulation in the near field. Figure 6.3 shows this figure of merit as a function of the wavelength for two different Co depths. The expected values for the magnetoplasmonic modulation are shown in Fig. 6.3. As it can be seen in this figure, the modulation depth ($4\Delta k_x^r d$) can be of about 2–3 % for $700 < \lambda < 900\,\text{nm}$ considering $d = L_{sp}$. It is relevant to notice that in order to have magnetic modulation of the interference intensity, we need to have two different counterpropagating SPPs. If we use an slit to launch an SPP, and a groove acting as a mirror, there would have been also interference between the two SPPs (the SPP traveling to the right, and the SPP that has reflected at the groove and travels to the left), but there would not be any magnetic modulation of this intensity. The phase $2\Delta k_x^r d$ that the SPP traveling to the left has obtained is the same that the one obtained when it was traveling to the right (before turning at the groove). In this case the expressions for the electromagnetic field would be the following (neglecting the k_z^i dependence):

$$
\begin{aligned}
\text{For } +\vec{M} &= +M \cdot \vec{y}: \\
E_{z,L}(+M) &= A \cdot e^{i(k_x^r + \Delta k_x^r)\cdot(x+d)}, \\
E_{z,R}(+M) &= A' \cdot e^{i(k_x^r + \Delta k_x^r)2d} e^{-i(k_x^r - \Delta k_x^r)(x-d)} \\
\text{For } -\vec{M} &= -M \cdot \vec{y}: \\
E_{z,L}(-M) &= A \cdot e^{i(k_x^r - \Delta k_x^r)\cdot(x+d)}, \\
E_{z,R}(-M) &= A' \cdot e^{i(k_x^r - \Delta k_x^r)2d} e^{-i(k_x^r + \Delta k_x^r)(x-d)},
\end{aligned}
\tag{6.7}
$$

that lead to the same intensity, therefore the MP intensity is zero.

6.1.2 Considering E_z and E_x

The previous analysis was a simplified calculation to easily derive the order of magnitude of the expected near field SPP intensity modulation induced by the application of an external magnetic field. Nevertheless, the actual situation in the near field is more complex. The next element that we are going to consider is that the SPP electromagnetic field has also an E_x component. In order to do that, we need to introduce not only the in-plane component of the SPP wavevector, but also the vertical component k_z (see Eq. 2.4). We are going to consider k_z as purely imaginary, $k_z = ik_z^i$, which means considering the SPP as an evanescent wave without any leakage radiation, a quite reasonable assumption, as can be deduced from Eq. 2.7. The in plane component of the wavevector will remain as purely real ($k_x = k_x{}^r$). The SPP field at the air side (the one accessible in a SNOM experiment) for the SPP launched at the left slit and propagating towards the right, when taking both E_x and E_z components into account is:

$$
\begin{aligned}
E_{z,L} &= A^d \cdot e^{ik_x{}^r\cdot(x+d)} e^{-k_{z,d}{}^i\cdot z} \\
E_{x,L} &= \frac{k_{z,d}{}^i}{k_x{}^r} A^d \cdot e^{ik_x{}^r\cdot(x+d)+i\pi/2} e^{-k_{z,d}{}^i\cdot z}
\end{aligned}
\tag{6.8}
$$

For the SPP launched at the right one, considering that $A^d = A^{d}$':

$$
\begin{aligned}
E_{z,R} &= A^d \cdot e^{-ik_x{}^r\cdot(x-d)} e^{-k_{z,d}{}^i\cdot z} \\
E_{x,R} &= \frac{k_{z,d}{}^i}{k_x{}^r} A^d \cdot e^{-ik_x{}^r\cdot(x-d)+i\pi/2} e^{-k_{z,d}{}^i\cdot z}
\end{aligned}
\tag{6.9}
$$

Resulting in a plasmonic intensity I^b:

$$
\begin{aligned}
I^b &= \left|\vec{E}_L + \vec{E}_R\right|^2 = \\
&= 2\left[1 + \left(\frac{k_{z,d}{}^i}{k_x{}^r}\right)^2\right] e^{-2k_{z,d}{}^i \cdot z}\left[A^2 + A^2\cos(2k_x{}^r \cdot x)\right] \\
&= \left[1 + \left(\frac{k_{z,d}{}^i}{k_x{}^r}\right)^2\right] e^{-2k_{z,d}{}^i \cdot z} \cdot I,
\end{aligned}
\tag{6.10}
$$

where I refers to the simpler expression obtained in Eq. 6.2.

To analyze the magnetic effect, we have to consider the magnetic modulation of k_z^i, that can be deduced from Eq. 2.7, and considering only linear terms, is defined as:

$$
\begin{aligned}
k_z(\vec{M}) &= k_z^0 \pm \Delta k_z \cdot m, \\
\text{with } \Delta k_z &= \frac{1}{2}\left[k_z(+\vec{M}) - k_z(-\vec{M})\right] \approx \frac{\Delta k_x k_x^0}{k_z^0}.
\end{aligned}
\tag{6.11}
$$

The magnetic field effect on the different components of the SPP wavevector depends on the relative direction between $\vec{M}$ and $\vec{k_x}$. The corresponding MP intensity is then:

$$
\begin{aligned}
I_{mp}^b &= \left[I^b(+M) - I^b(-M)\right] \\
&\approx -\left[1 + \left(\frac{k_z^{i,2} - \Delta k_z^{i,2}}{k_x^{r,2} - \Delta k_x^{r,2}}\right)\right] \cdot 8A^2 e^{-2k_z{}^i z} \Delta k_x^r d \sin(2k_x x) \\
&\approx \left[1 + \left(\frac{k_z^{i,2} - \Delta k_z^{i,2}}{k_x^{r,2} - \Delta k_x^{r,2}}\right)\right] \cdot I_{mp} \\
&\approx \left[1 + \left(\frac{k_z^{i,2}}{k_x^{r,2}}\right)\right] \cdot I_{mp} \approx I_{mp},
\end{aligned}
\tag{6.12}
$$

where we have taken the following approximations: $2\Delta k_x^r d \ll 1$, $\Delta k_x^r \ll k_x^r$, $\Delta k_{z,d}^i \ll k_{z,d}^i$, and $k_{z,d}^{i,2} \ll k_x^{r,2}$.

Finally, the obtained expression is equivalent to Eq. 6.5. Table 6.1 presents the ratio of E_z and E_x, equivalent to the ratio $k_x^r / k_{z,d}^i$, which shows that disregarding E_x is not a bad approximation, mainly for longer wavelengths.

6.1.3 Introducing Propagation Losses, k^i, Leakage Radiation k_z^r, and Their Modulations

Let us now consider losses, meaning to consider the imaginary part of the SPP wavevector, k_x^i, and its magnetic modulation. Besides, k_z will be completely complex, without any approximation, and both components will be modulated (although k_z^r is going to be really negligible). Taking into account the results of the previous

Table 6.1 Calculated values for k_x and k_z and ratio between the two components of the electromagnetic field of the plasmon E_z/E_x for $h = 15$ nm

λ_0 (nm)	k_x^r (μm^{-1})	k_x^i (μm^{-1})	k_z^r (μm^{-1})	k_z^i (μm^{-1})	$E_z/E_x = k_x^r/k_z^i$ (μm^{-1})
540	12.55	0.60	−1.53	4.92	2.55
633	10.39	0.099	−0.34	3.06	3.39
690	9.42	0.049	−0.19	2.41	3.91
785	8.19	0.023	−0.11	1.74	4.70
980	6.50	0.009	−0.06	1.05	6.18

subsection, we will consider only E_z so that:

$$\begin{aligned} k_x &= k_x^r + ik_x^i \\ k_x(\vec{M}) &= k_x^0 \pm \Delta k_x \cdot m \text{ with } \Delta k_x = \Delta k_x^r + i\Delta k_x^i \\ k_z &= k_z^r + ik_z^i \\ k_z(\vec{M}) &= k_z^0 \pm \Delta k_z \cdot m \text{ with } \Delta k_z = \Delta k_z^r + i\Delta k_z^i, \end{aligned} \tag{6.13}$$

being the expression of the electromagnetic field of the left and right slit as:

$$\begin{aligned} E_{z,L} &= A \cdot e^{ik_x{}^r\cdot(x+d)} e^{ik_z^r\cdot z} e^{-k_z{}^i\cdot z} e^{-k_x^i\cdot(x+d)} \\ E_{z,R} &= A' \cdot e^{-ik_x{}^r\cdot(x-d)} e^{ik_z^r\cdot z} e^{-k_z{}^i\cdot z} e^{k_x^i\cdot(x-d)} \end{aligned} \tag{6.14}$$

This results in the following interference intensity when there is no magnetic field applied:

$$\begin{aligned} I^c &= e^{-2k_{z,d}^i z} e^{-2k_x^i d} \left[A^2 e^{-2k_x^i x} + A'^2 e^{2k_x^i x} + 2A'A\cos\left(2k_x^r x\right)\right], \\ &\cong 2A^2 \cdot e^{-2k_{z,d}^i z} e^{-2k_x^i d}\left[\cosh\left(2k_x^i x\right) + \cos\left(2k_x^r x\right)\right], \\ &\quad \text{for } A = A' \end{aligned} \tag{6.15}$$

A reasonable expression of the MP intensity could be obtained under some approximations, such as taking the first order approximation for smaller enough products $\Delta k_z^{r,i} z$, $\Delta k_x{}^i x$, and $\Delta k_x{}^{i,r} d$:

$$\begin{aligned} I_{mp}^c \approx -8A^2 e^{-2k_z^i z} e^{-2k_x^i d} \Big[& \Delta k_x^i x \left(\cosh(2k_x^i x) + \cos(2k_x^r x)\right) \\ & + (\Delta k_x^r d + \Delta k_z^r z)\sin(2k_x^r x) \\ & - \left(\Delta k_x^i d + \Delta k_z^i z\right)\sinh(2k_x^i x)\Big] \end{aligned} \tag{6.16}$$

Table 6.2 Calculated values for the MP modulation of both k_x and k_z for Au/Co/Au trilayers

λ_0 (nm)	$h = 15\,\text{nm}$			$h = 10\,\text{nm}$		
	$\Delta k_x^r \times 10^{-4}$ (μm^{-1})	$\Delta k_x^i \times 10^{-4}$ (μm^{-1})	$\Delta k_z^i \times 10^{-4}$ (μm^{-1})	$\Delta k_x^r \times 10^{-4}$ (μm^{-1})	$\Delta k_x^i \times 10^{-4}$ (μm^{-1})	$\Delta k_z^i \times 10^{-4}$ (μm^{-1})
540	17.8	9.85	47.8	22.7	11.3	62
633	6.8	−0.45	22.7	9.2	−0.55	31.1
690	4.15	−1.05	15.8	5.85	−1.35	22.6
785	2.4	−1.1	10.7	3.35	−1.05	15.3
980	1.5	−1.05	8.9	2.2	−1.5	13

Considering the values of all the parameters involved (indicated in Table 6.2), the approximations are quite reasonable. In certain conditions (small x and long L_{sp}), it could also be considered $k_x^i x \cong 0$, thus $\cosh(2k_x^i x) \cong 1$ and $\sinh(2k_x^i x) \cong 0$, resulting in:

$$I_{mp}^c \approx -8A^2 e^{-2k_z^i z} e^{-2k_x^i d} \Big[\Delta k_x^i x + \sqrt{(\Delta k_x{}^i x)^2 + (\Delta k_x{}^r d)^2} \sin(2k_x^r x + \alpha) \Big], \tag{6.17}$$

where $\tan\alpha \approx \alpha \approx \frac{\Delta k_x^i x}{\Delta k_x^r d}$. Similarly to the far field case (Chap. 3), the modulation of the imaginary part of the in plane SPP wavevector induces a phase shift in the near field MP interferogram. As Eq. 6.17 shows, the expression is the same as for the simple case (Eq. 6.5) except for the corresponding exponential decay, an added offset and an extra phase shift (α). Figure 6.4 shows the NF plasmonic and MP intensities calculated when taking the losses into account, and when they are discarded. As it can be seen, the offset depends on the position between the slits, and although it is

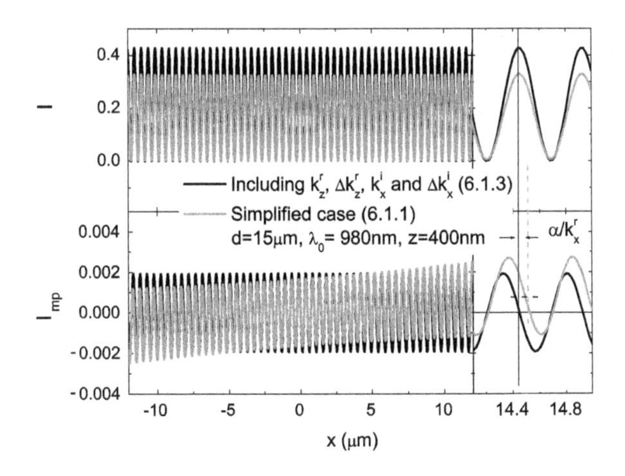

Fig. 6.4 NF plasmonic and MP intensities when plasmon propagation losses are taken into account or discarded. The *panels of the right* show those intensities for a smaller range of x in order to appreciate the phase shift between both signals when the losses are considered. Be aware of the change in the scale for x axis

not shown, it depends also on the wavelength. It is a straight line that increases with the position between the slits x, so that the sinusoidal MP pattern is not centered at $I_{mp} = 0$ (except at $x = 0$). The phase shift, on the other hand, is small and it is more difficult to distinguish. This is shown in the panels on the right, where we can see that a maximum of the plasmonic intensity matches with a zero (the middle of the amplitude signal) of the MP intensity for the simplified case, while for the complete situation it is slightly shifted. If we calculate the contrast of I_{mp}/I for both graphs of Fig. 6.4, we can see that for the simplified case the contrast is about $9 \cdot 10^{-3}$, while for the analysis including losses it can be of approximately $1 \cdot 10^{-2}$ for $x = 15\,\mu\text{m}$. This means that the effect of considering the losses modulation is about a 10 % of the total signal at big x values. Therefore, the simplified situation of Sect. 6.1.1 seems appropriate regarding the ratio between the simplicity of the obtained expressions compared with the introduced error. Moreover, although Δk_z^i is of the order of $\Delta k_x^{r,i}$, being z much smaller than x (about 400 nm over several microns) in the SNOM measurements makes the contribution of Δk_z^i negligible in this configuration. The main difference in the obtained results between the calculations including losses and the simplified situation of Sect. 6.1.1 is the asymmetry in the MP intensity. This is due to the asymmetry of the magnetic field effect on both SPPs, the right and the left ones, transmitted to the amplitude of the signal by k_x^i.

6.1.4 Measuring Fluorescence (E^4)

There are SNOMs that using a fiber tip measure the intensity, but we want to consider also SNOM experiments where a fluorescent particle is placed at the SNOM tip [5]. In this case the detected signal is the 4th power of the electromagnetic field, instead of the intensity. So now we will consider the simplest case described in Sect. 6.1.1, but analyzing E_z^4 instead of E_z^2 to check whether there would be any important difference in the plasmonic and MP intensities.

From Eq. 6.1, the total E_z component of the electromagnetic field is:

$$E_z = Ae^{-k_z^i z}\left[e^{ik_x^r(x+d)} + e^{-ik_x^r(x-d)}\right] \approx 2Ae^{-k_z^i z}e^{ik_x^r d}\cos(k_x^r x), \qquad (6.18)$$

where we assume $A = A'$. Hence the purely plasmonic signal detected by fluorescence (S) is:

$$S = |\vec{E}|^4 = 2A^4 e^{-4k_z^i z}\left[3 + \cos(4k_x^r x) + 4\cos(2k_x^r x)\right], \qquad (6.19)$$

The MP signal is:

$$S_{mp} = S(+M) - S(-M) \qquad (6.20)$$

which, for the simplest case that takes into account only Δk_x^r, becomes:

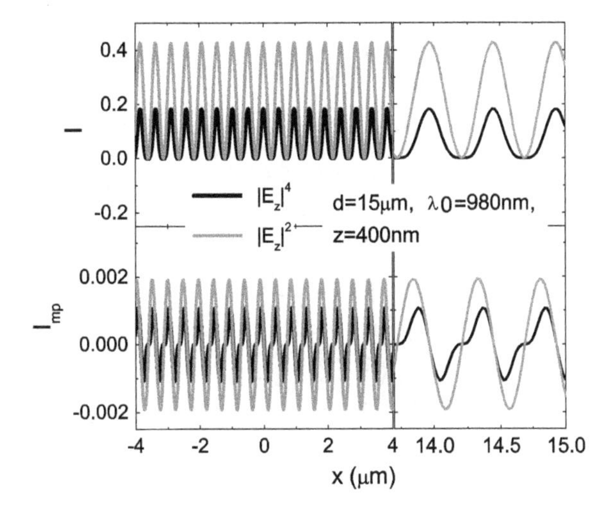

Fig. 6.5 Difference between considering measuring with the fluorescent particle (E_z^4) or an optical fiber tip (E_z^2) for $\lambda_0 = 980\,\text{nm}$, $z = 400\,\text{nm}$ and a distance between slits of $d = 15\,\mu\text{m}$, being $A = 0.5$

$$S_{mp} = -16A^4 \Delta k_x^r d e^{-4k_z^i z} \left[2\sin(2k_x^r x) + \sin(4k_x^r x)\right] \tag{6.21}$$

The difference between this fluorescence configuration (E_z^4) and the case of measuring intensity (E_z^2) can be seen in Fig. 6.5. The right part of the graph shows a zoom of the oscillations just to appreciate the difference in shape between both situations. As it can be seen, the shape of the plasmonic term is in both cases a periodic function with the same period ($\lambda_{sp}/2$). For the MP term, the fluorescence signal is also perfectly periodic, but the sinusoidal shape is slightly distorted. The most important point is that the ratio between the MP and the plasmonic contrasts is a little bit larger for the fluorescence situation than in the intensity case, a 30 % larger, as can be deduced from Fig. 6.5.

6.1.5 Integrating the Volume of the Particle

In the previous subsection we have introduced the effect of the fluorescent particle as a non-linear effect, considering that we measure the fourth power of the field instead of the square. Another aspect that we should take into account is the particle or tip size, therefore integrating the corresponding power of the field over all the particle volume. This is relevant for both measuring with fluorescent particles at the tip or using fiber tips.

We have therefore calculated the NF intensity ($|E|^2$) for the simplified case (Sect. 6.1.1), comparing the point tip situation with the volume integration case. As it can be seen in Fig. 6.6, when the integration over the tip size is done, a huge decrease of the total signal occurs (the total signal decreases due to the size of the molecule, taken here as 0.4 μm, and the contrast decreases due to the average). However, when considering the ratio of the contrasts of the MP signal and the plasmonic

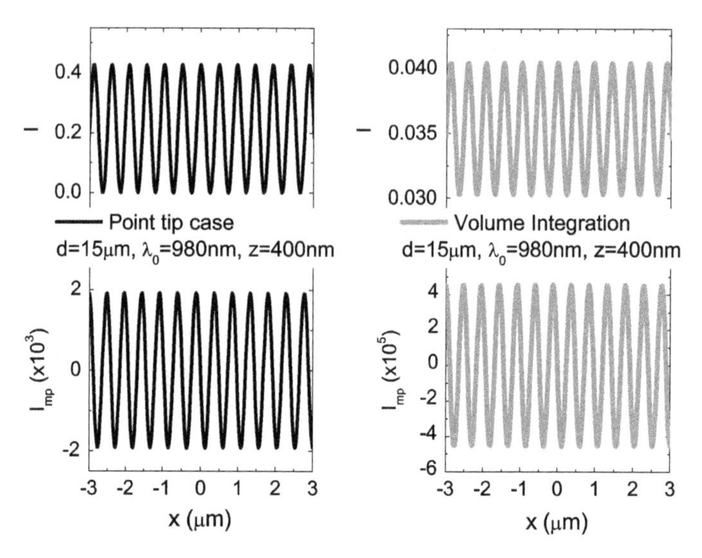

Fig. 6.6 Difference in the plasmonic and MP intensity when considering a punctual tip (*left*) or a finite size tip (*right*). The analyzed signal in this case is the intensity ($|E|^2$)

one, the result is the same with and without integration. Although it is not shown, we have seen that if we consider the fact that we measure the fourth power of the electromagnetic field ($|E|^4$) integrated all over the particle size, the results are similar. When integrating over the volume of the particle, the signal is dramatically reduced, which would worsen the measurements, but the ratio of the contrasts between the MP signal and the plasmonic one are the same that without integrating the volume of the particle.

6.1.6 *Summary*

Summarizing, magnetically induced NF intensity modulations ($4\Delta k_x \cdot d$) of around 2–3 % are expected in the case of two counterpropagating SPPs in Au/Co/Au trilayers at $\lambda_0 = 950$ nm, associated with the modification of the SPP wavevector. This modulation could in principle be detected with a SNOM system if the signal-to-noise ratio is low enough.

6.2 Single SPP. Evaluation of Δk_z and Δk_x^i

In the near field, there is not any need to have interferences to be able to see SPPs. In fact, there is not any physical effect that forbids measuring the electromagnetic field of SPP by SNOM. Therefore in this section, another configuration is analyzed,

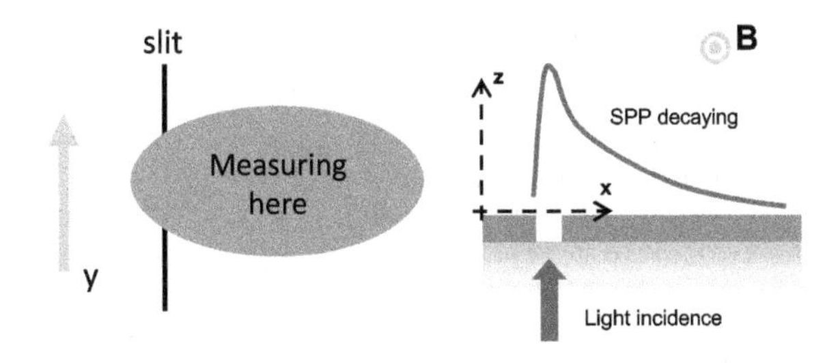

Fig. 6.7 Sketch of the configuration for the proposed single SPP experiment. There is only one slit to launch the SPP, and we can see only the SPP decaying field, without any interference

trying to take advantage of the modulations in the imaginary part and in the vertical component of the wavevector, k_z, with the aim of determining whether they could be measured or not. In order to have access to Δk_x^i and Δk_z, measurements with an isolated slit can be done (see Fig. 6.7). We are going to calculate directly the signal collected with the optical fiber tip and the fluorescent particle. We should have in mind that in this configuration there are not interferences, and as a consequence there will be no contribution of k_x^r. In this case, at a given point (x_0, z_0), the z component of the SPP field is given by:

$$E_z = A \cdot e^{ik_x x_0} e^{-k_z^i \cdot z_0}, \text{ so } |E_z|^2 = A^2 e^{-2k_x^i \cdot x_0} e^{-2k_z^i \cdot z_0}, \text{ and}$$
$$|E_z|^4 = A^4 e^{-4k_x^i \cdot x_0} e^{-4k_z^i \cdot z_0} \tag{6.22}$$

When we apply the magnetic field and consider the difference between the signal with the magnetic field in opposite directions divided by the intensity with no magnetic field applied, it yields:

$$\begin{aligned} \frac{|E_z(+M)|^2 - |E_z(-M)|^2}{|E_z(M=0)|^2} &= -2\sinh(2\Delta k_x^i x_0 + 2\Delta k_z^i z_0) \\ &\approx -4(\Delta k_x^i x_0 + \Delta k_z^i z_0), \\ \frac{|E_z(+M)|^4 - |E_z(-M)|^4}{|E_z(M=0)|^4} &= -2\sinh(4\Delta k_x^i x_0 + 4\Delta k_z^i z_0) \\ &\approx -8(\Delta k_x^i x_0 + \Delta k_z^i z_0); \end{aligned} \tag{6.23}$$

The most interesting aspect that this expression shows is that, in principle, the contribution of Δk_x^i and Δk_z^i could be distinguished by making profiles along both x and z directions. The theoretical values of Δk_x^i and Δk_z^i obtained with the experimentally determined optical and MO constants can be found in Table 6.2. For those values, the obtained modulation ratios are around 1–4 % depending on x_0 and z_0, so that this configuration is promising for experimental detection. Since the modulation is linear with both x_0 and z_0, if we are above the noise level, profiles along x and z directions could provide the experimental value of Δk_x^i and Δk_z^i. As Fig. 6.8c shows, the ratio between the modulated electromagnetic field and the plasmonic intensity increases as we go further from the slit, but due to the evanescent nature of the

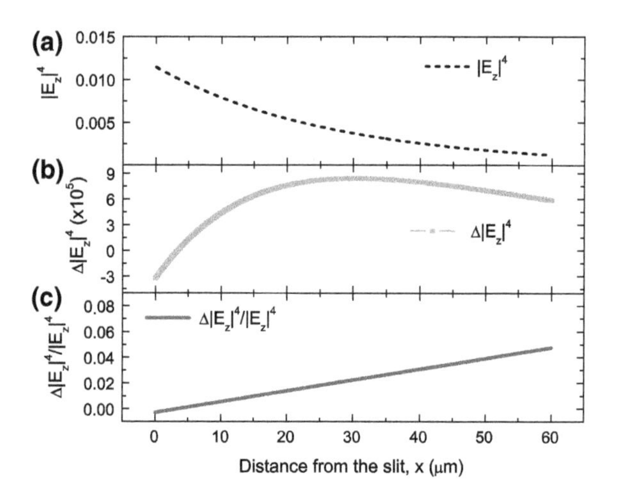

Fig. 6.8 **a** NF Fluorescent intensity ($|E|^4$) for a single SPP travelling in a Au/Co/Au trilayer, **b** magnetic modulation of the fluorescent intensity ($\Delta|E_{sp}|^4$), **c** ratio between the two intensities. Results are obtained for a system with $h = 15$ nm and at $\lambda_0 = 980$ nm, and $z = 400$ nm. It can be seen that there are no fringes, or any resemblance of interference, and despite the small signal values, the ratio is very promising

plasmon, the pure plasmonic signal (Fig. 6.8a) is also decaying, thus worsening the possible measurement conditions. In fact, the value of the initial signal (see Fig. 6.8) is quite small, and because of this small value of the intensity of the electromagnetic field, getting a signal from SPPs with a SNOM it is usually done through an interferometric technique [5, 9, 11, 12] or a spectroscopic one [13–16]. This implies that even though in this geometry we could theoretically detect modulation magnitudes that in other configurations are not possible (Δk_x^i and Δk_z^i), and although the ratio is quite good, the signal-to-noise ratio could be too low to make this SNOM experiment practicable.

6.3 Interference of SPP with Radiative Light

Finally, let us discuss the configuration proposed in Ref. [11], where the SPP is excited by lateral incidence on a slit, and introduce the effect of an applied magnetic field (Fig. 6.9). In this case, interference between the radiative light (incident and reflected) and the surface plasmon is observed. Regarding the SPP, the magnetic field would modify k_x, as above. For the radiative light, on the other hand, k_0 is not modified but the reflectivity coefficient, r, is. This is the standard transverse Kerr effect in magneto-optical materials [17–19].

Again, the idea is to apply an alternating magnetic field and to detect variations of intensity ΔI at each point synchronous with the applied magnetic field (as we have done in the first section of this chapter). This variation of intensity could then be calculated following the same procedure as in the case of parallel slits, taking the equations presented in Ref. [5] and introducing the effect of the magnetic field. Here it is very important to take into account that we have two contributions: Δk of the SPP, but also Kerr effect for the radiative light. This last contribution has been explained

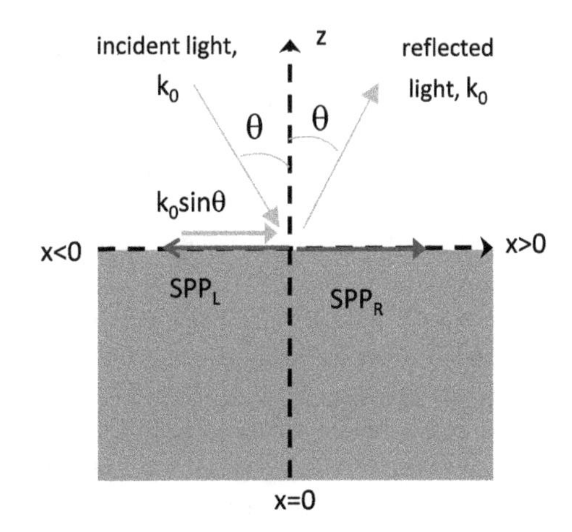

Fig. 6.9 Diagram of the waves that interfere in this configuration. There are two plasmons that propagate in opposite directions and the incident and reflected waves. Interference of each SPP and the incident and reflected waves (radiative light) takes place

in detail in Sects. 2.3 and A.2.2. The transverse Kerr effect can be formulated by a reflectivity coefficient that depends on the sample magnetization: $r = r^0 + a \cdot m$, where a is a parameter that depends on the layer structure and m is the magnetization of the sample, which is 1 or -1 since we are saturating the sample.

It would be relevant to find a way to separate the contributions of the modification of k_x and the Kerr effect. One idea would be to modify the angle of the excitation light, as the Kerr effect depends on the angle of incidence (it decreases when the angle of incidence θ decreases and it is actually zero for $\theta_{inc} = 0$). In the case we could work at $\theta_{inc} = 0$, normally to the slit, a simplified estimation of ΔI (working with E_{sp}^2, $\Delta k_x = \Delta k_x^r$, and $\Delta k_z = \Delta k_z^i$) can be made; the results are the same as those obtained in the previous section (Sect. 6.2).

6.3.1 45° Incidence

At not normal incidence, for example at $\theta = 45°$, the Kerr effect has to be considered. In this case, given the complex situation, we have only considered E_z, as in Ref. [5]. The analysis has been carried out numerically, considering that we measure intensity (not fluorescence), and that all the parameters are complex. In order to be able to compare this situation with the previous ones, the simulations are made for the sample with 15 nm Au on top. An sketch of this geometry can be seen in Fig. 6.9. The expressions for the fields at both sides of the slit are then [9]:

$$
\begin{aligned}
E_z^R(x,z) = &- H_0 \sin(\theta) e^{ik_0 \sin(\theta) x} \left[e^{-ik_0 \cos(\theta) z} + r \cdot e^{ik_0 \cos(\theta) \cdot z} \right] \\
&- A^R \frac{k_x}{k_0} e^{ik_x x} e^{ik_z z}, \qquad (6.24) \\
&\text{for } x > 0 \text{ and being } H_0 = 1 \text{ and } A^R = 0.6\,[5]
\end{aligned}
$$

$$E_z^L(x,z) = -\, H_0 \sin(\theta) e^{ik_0 \sin(\theta) x} \left[e^{-ik_0 \cos(\theta) z} + r \cdot e^{ik_0 \cos(\theta) \cdot z} \right] + A^L \frac{k_x}{k_0} e^{-ik_x x} e^{ik_z z}, \tag{6.25}$$

for $x < 0$ and being $H_0 = 1$ and $A^L = 0.45$ [5].

where H_0 is the amplitude of the incident plane wave, while A^R and A^L represent the amplitudes of the SPPs launched at both sides of the slit.

Due to the oblique incidence, the amplitudes of the SPP launched at both sides (A^R and A^L) are different [9], as shown in Eqs. 6.24 and 6.25. Moreover, the main difference between the two sides of the slit is that, given the different sign of k_x regarding k_0, there is a noticeable difference in the period of the interferences [9]: for $x < 0$ the period is $\frac{2\pi}{k_0 \sin\theta + k_x^r}$, which is about $\lambda/2$, while for the $x > 0$ case is $\frac{2\pi}{k_0 \sin\theta - k_x^r}$ which is about 3λ. This last period is so large that we have focused ourselves to the $x < 0$ situation, since is the one most likely to succeed in an experiment, because an interference with a smaller period is easier to identify. When we apply the magnetic field, both r and k_x are going to change, such as. And it is important to notice at which side of the slit we are, since for $x < 0$, due to the opposing signs, the effect of the magnetic field will be the opposite for the incidence light and for the SPP. Below, the effect of the magnetic field is written for the $x < 0$ side:

$$\begin{aligned} E_z^L(+B) &= -H_0 \sin(\theta) e^{ik_0 \sin(\theta) x} \left[e^{-ik_0 \cos\theta z} + (r^0 + a) \cdot e^{ik_0 \cos(\theta) \cdot z} \right] \\ &\quad + A^L \frac{(k_x - \Delta k_x)}{k_0} e^{-i(k_x - \Delta k_x) x} e^{i(k_z - \Delta k_z) z} \\ E_z^L(-B) &= -H_0 \sin(\theta) e^{ik_0 \sin(\theta) x} \left[e^{-ik_0 \cos\theta z} + (r^0 - a) \cdot e^{ik_0 \cos(\theta) \cdot z} \right] \\ &\quad + A^L \frac{(k_x + \Delta k_x)}{k_0} e^{-i(k_x + \Delta k_x) x} e^{i(k_z + \Delta k_z) z}, \end{aligned} \tag{6.26}$$

where a is the magnetic effect on the r coefficient and Δk_x and Δk_z are the magnetic modulation of the x and z component of the SPP wavevector considering that we work at saturation values. Table 6.3 shows the values of r, a, k_x^r and Δk_x^r for the analyzed wavelength. The relative change induced by the magnetic field is almost one order of magnitude higher for the Fresnel coefficient r than for k_x^r; however, the difference is not so huge to discard the second term of Eq. 6.26. An approximation to weight the different contributions can be obtained by splitting the magnetic modulation of E_z^L in the following way:

Table 6.3 Expected values for the Fresnel coefficient r and its magnetic modulation a, together with the values of k_x^r and Δk_x^r at $\lambda_0 = 980$ nm and for a Co depth of 15 nm and incidence at $\theta = 45°$

Wavelength	r	a ($\times 10^{-4}$)	k_x^r (μm^{-1})	Δk_x^r ($\times 10^{-4}\ \mu m^{-1}$)
980 nm	0.967	4.1	6.50	1.5

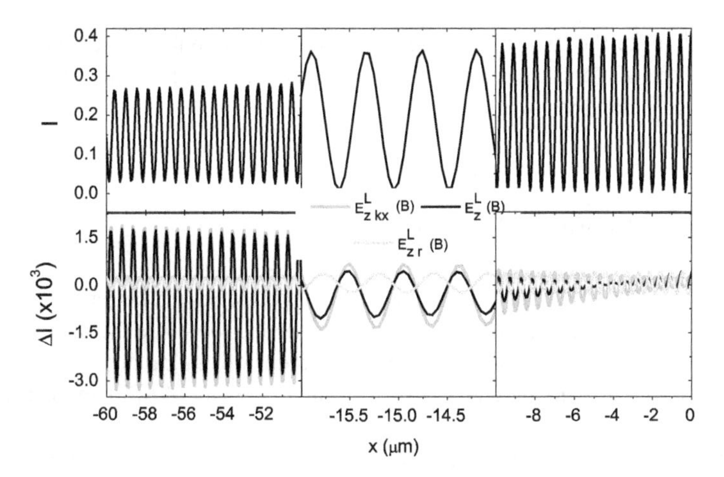

Fig. 6.10 Intensity and MP intensity for the case of a SPP that propagates along a surface in the near field and interferes with incident and reflected radiative light traveling in the opposite direction. Three situations are considered: that only the plasmon related parameters k_x and k_z depend on B ($E^L_{z\,k_x}(B)$); that the three parameters r, k_x and k_z depend on the magnetic field B ($E^L_z(B)$); and finally that only the Fresnel coefficient depends on the magnetic field B ($E^L_{z\,r}(B)$). The *different panels* show different x scales. These results are for a system with $h = 15\,\text{nm}$ Au and at $\lambda_0 = 980\,\text{nm}$ and $z = 400\,\text{nm}$

$$
\begin{aligned}
E^L_z(B) = &- H_0 \sin(\theta) e^{ik_0 sin(\theta)x} \left[e^{-ik_0 \cos\theta z} + r(B) \cdot e^{ik_0 \cos(\theta)\cdot z}\right] \\
&+ A^L \frac{k_x(B)}{k_0} e^{-ik_x(B)x} e^{ik_z(B)z} \\
E^L_{z\,r}(B) = &- H_0 \sin(\theta) e^{ik_0 \sin(\theta)x} \left[e^{-ik_0 \cos\theta z} + r(B) \cdot e^{ik_0 cos(\theta)\cdot z}\right] \\
&+ A^L \frac{k_x}{k_0} e^{-ik_x x} e^{ik_z z} \\
E^L_{z\,kx}(B) = &- H_0 \sin(\theta) e^{ik_0 \sin(\theta)x} \left[e^{-ik_0 \cos\theta z} + r \cdot e^{ik_0 \cos(\theta)\cdot z}\right] \\
&+ A^L \frac{k_x(B)}{k_0} e^{-ik_x(B)x} e^{ik_z(B)z},
\end{aligned}
\tag{6.27}
$$

where for $E^L_{z\,kx}(B)$ the only parameters that depend on the magnetic field are k_x and k_z, while for the $E^L_{z\,r}(B)$ case, only r is going to depend on the magnetic field B. As it can be seen in Fig. 6.10, the main contribution to the MP intensity (ΔI) is due to the modulation of the plasmon (except at small distances), contrary to what could be expected from Table 6.3. Although it is not shown, for larger angles the effect of r becomes more relevant, but it can still be neglected. The ratio of the contrasts at $x = 15\ \mu\text{m}$ is about $\frac{\Delta I_{contrast}}{I_{contrast}} = 5 \cdot 10^{-3}$, which is smaller than that for the two counterpropagating SPPs configuration, although for larger distances ($x = 60\ \mu\text{m}$), this ratio can be of about 2 %, and it increases with the distance. However, the initial plasmonic signal decreases as we go far away from the slit, so even if the ratio could be bigger than for the ‘two parallel slits’ situation, the signal-to-noise ratio is likely to worsen strongly and we do not think that this is an experimental configuration more adequate than the ‘two parallel slits’ case.

6.4 Experimental Implementation

As we have previously seen, the configuration that provides the larger magnetic field modulated intensity, and also large intensity values and therefore the better expected signal to noise ratio is the one where two counterpropagating SPPs interfere. We have tried to measure the magnetically modulated intensity of SPPs in the near field in collaboration with Lionel Aigouy from the ESPCI in France. In this section we are going to describe the implementation of this NF modulation experiment.

Near field optics involves fields that are spatially very confined. This implies the need of a detection technique that can be located very close (tens of nanometers) to the surface. The invention of the SNOM has allowed this. Some concepts of the SNOM technique will be explained here, in order to give a global idea of this technique and the possibilities it can provide [20]. SNOM is a local probe microscope similar to the more common scanning tunneling microscope STM [21] or AFM [22]. It is based on a tip that is placed very close to the surface of the sample and is controlled with a piezoelectric stage and a laser. This tip can work at a fixed distance from the sample, or following the topography of the sample. This topography, or working distance, is used as a reference or feedback. The SNOM microscope collects the electromagnetic field close to the sample and in this way it maps its spatial distribution. However, due to the interaction between the tip and the sample, and/or due to background light, the physical interpretation of the obtained maps is often not trivial. One of the advantages of SNOM is that it makes it possible to detect evanescent waves such as SPPs [11–13, 15] or guided modes [23], or the local distribution of quantum dots’ emission [24]. In SPP experiments the working distance is smaller than λ_0 ($z = 400\,\mathrm{nm}$ in [9] for $\lambda_0 = 975\,\mathrm{nm}$). This implies that resolutions of about tens of nanometers are usually achieved [25].

There are several configurations for SNOM microscopes, such as those working in reflection or transmission, in collection or illumination, or using a fiber tip with an aperture or without it (apertureless). Besides, there is also the possibility of working with a fluorescent particle [10, 20, 26] instead of a fiber; or working with a metal coated fibre. In our case, we have collaborated with a group whose SNOM microscope uses an Erbium doped fluorescent particle placed in a Tungsten tip [10].

The configuration we propose is presented in Ref. [5], where the SNOM works in collection mode. To access different wavelengths several particles can be used, each one with its own fluorescent excitation wavelength. For the erbium doped particle, the most suitable wavelength is 980 nm. The main characteristic of using fluorescent particles as a tip is that the collected signal is that emitted by the particle, not by the sample, and it is proportional to the fourth power of the sample electromagnetic field. Moreover, the particle size (about 400 nm [5] of diameter) must be taken into

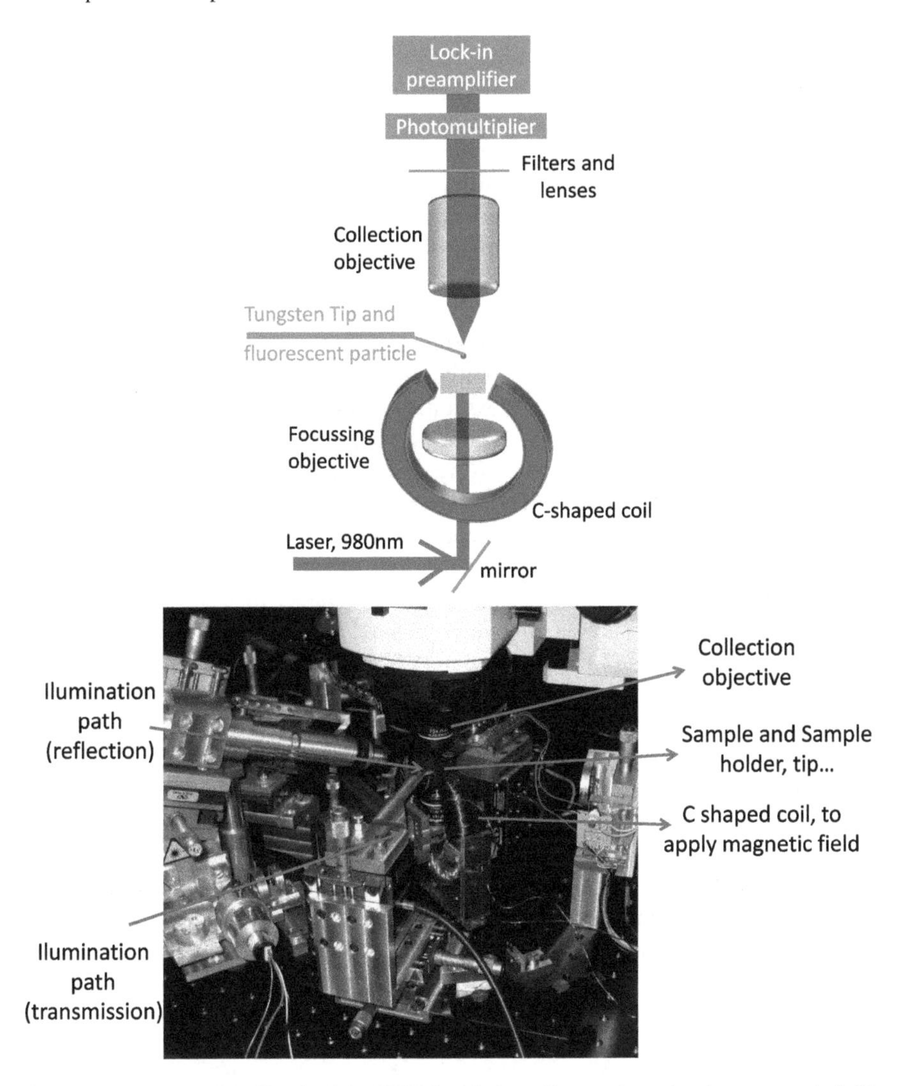

Fig. 6.11 (*Upper image*) Sketch of the SNOM with the coil to apply an in-plane magnetic field. (*Lower image*) Photograph of the actual SNOM setup used to make the experiments. It shows a transmission and collection SNOM with a fluorescent particle at the tip

account, being the complete collected signal the integral over the volume of the particle of the fourth power of the field.

This is basically how the used SNOM works. However, for our measurements, we need to apply a magnetic field to the sample, so a coil is located under the sample holder. In Fig. 6.11, a sketch of the used SNOM, together with a photograph of the experimental setup, can be seen.

With a p-polarized laser illuminating from below and at normal incidence to avoid the Faraday effect, we generate four SPP, two at each slit (Fig. 6.1). The two central SPPs will interfere all over the space between the two slits, creating a sinusoidal pattern (Eq. 6.2). Therefore scanning with the SNOM this central area between the slits, we are able to see a series of maxima and minima. When we apply a magnetic field parallel to the slits, the wavevector of the two SPPs is going to change, resulting in a shift of this interference pattern and in a variation of the intensity. This MP signal can be measured applying an alternating magnetic field of about 20 mT, which is high enough to saturate the sample [1], and using a lock-in amplifier to detect for each x position the change in intensity synchronous with the magnetic field. Therefore I_{mp} is the component in the lock-in amplifier at the frequency of the magnetic field and $I(M = 0)$ is the DC component. As it has been shown in previous sections, the MP intensity is proportional to the product $4\Delta kd$, being $2d$ the distance between slits, a known and selected parameter. Nevertheless, d cannot be any distance, since it must be related to the propagation distance of the plasmons L_{sp}, as is explained in Chap. 3. Since several wavelengths could be used, interferometers with several distances between the slits were fabricated, varying from 5 to 30 µm.

Based on Ref. [6], and given the expected values obtained in the previous sections, we performed a series of measurements of this MP intensity in collaboration with Aigouy's laboratory at the LPEM-CNRS in Paris. Unfortunately, it was not possible to draw any conclusions due to the noise level. This was probably due to the fact that the modulation obtained in previous sections is referred to the plasmonic sinusoidal **contrast**, not to the total plasmonic intensity I (in the total plasmonic intensity the laser shape and background light influence is large). Currently our collaborators are working on increasing the SNOM signal-to-noise ratio. Another fact to consider is that we need to measure differences of about 4 % in the oscillations of the interference, that already need a lock-in technique. Since there is not enough signal with the uncovered interferometers, a further step could be using PMMA covered interferometers. However, this has to be carefully considered since the tip would be at a larger distance from the metal/dielectric interface, and there will be less detected signal. Moreover, the measurements will be more difficult as the tip can get stuck in PMMA.

6.5 Conclusions

We have analyzed three possible configurations to obtain magnetic modulation of SPP in the near field. However, we have seen that there are two of them specially interesting, due to their simple experimental implementation, to the physics involved, and to the possibility of really measuring the modulated intensity. One of them consist of two counterpropagating SPPs, being the modulation depth proportional to $4\Delta k_x \times d$, that could achieve values of about 2–3 % for an interferometer with an upper Au layer of 15 nm at $\lambda_0 = 700$–900 nm and $d = L_{sp}$. On the other hand, a single SPP (without any interference) could provide information about parameters that couldn't be obtained otherwise, such as Δk_x^i or Δk_z^i, which also provides reasonable modulation depths. The first configuration has been performed experimentally at the ESPCI in France in collaboration with Lionel Aigouy. Experiments were done with interferometers consisting of two parallel slits engraved on Au/Co/Au trilayers with 10 and 15 nm of gold in the upper layer. Different distances between the slits were tried. However, the signals were too low and we could not confirm that we measured magnetically modulated intensity.

References

1. V.V. Temnov, G. Armelles, U. Woggon, D. Guzatov, A. Cebollada, A. Garcia-Martin, J.M. Garcia-Martin, T. Thomay, A. Leitenstorfer, R. Bratschitsch, Nat. Photonics **4**, 107–111 (2010)
2. D. Martin-Becerra, J.B. Gonzalez-Diaz, V. Temnov, A. Cebollada, G. Armelles, T. Thomay, A. Leitenstorfer, R. Bratschitsch, A. Garcia-Martin, M.U. Gonzalez, Appl. Phys. Lett. **97**, 183114 (2010)
3. D. Martin-Becerra, V.V. Temnov, T. Thomay, A. Leitenstorfer, R. Bratschitsch, G. Armelles, A. Garcia-Martin, M.U. Gonzalez, Phys. Rev. B **86**, 035118 (2012)
4. H.W. Kihm, K.G. Lee, D.S. Kim, J.H. Kang, Q.-H. Park, Appl. Phys. Lett. **92**, 051115 (2008)
5. L. Aigouy, P. Lalanne, J.P. Hugonin, G. Julié, V. Mathet, M. Mortier, Phys. Rev. Lett. **98**, 153902 (2007)
6. J.-C. Weeber, Y. Lacroute, A. Dereux, E. Devaux, T. Ebbesen, C. Girard, M.U. González, A.-L. Baudrion, Phys. Rev. B **70**, 235406 (2004)
7. P. Lalanne, J.P. Hugonin, J.C. Rodier, J. Opt. Soc. Am. A **23**, 1608–1615 (2006)
8. G. Gay, O. Alloschery, B. Viaris de Lesegno, C. O'Dwyer, H.J. Weiner, J. Lezec, Nature **2**, 262–267 (2006)
9. B. Wang, L. Aigouy, E. Bourhis, J. Gierak, J.P. Hugonin, P. Lalanne, Appl. Phys. Lett. **94**, 011114 (2009)
10. L. Aigouy, Y. De Wilde, M. Mortier, Appl. Phys. Lett. **83**, 147–149 (2003)
11. H. Gao, J. Henzie, T.W. Odom, Nano Lett. **6**, 2104–2108 (2006)
12. J. Chen, M. Badioli, P. Alonso-Gonzalez, S. Thongrattanasiri, F. Huth, J. Osmond, M. Spasenovic, A. Centeno, A. Pesquera, P. Godignon, A. Zurutuza-Elorza, N. Camara, F.J. Garcia de Abajo, R. Hillenbrand, F.H.L. Koppens, Nature **487**, 77–81 (2012)
13. T. Klar, M. Perner, S. Grosse, G. Von Plessen, W. Spirkl, J. Feldmann, Phys. Rev. Lett. **80**, 4249–4252 (1998)
14. S. Benrezzak, P.M. Adam, J.L. Bijeon, P. Royer, Surf. Sci. **491**, 195–207 (2001)
15. Z. Fang, X. Zhang, D. Liu, X. Zhu, Appl. Phys. Lett. **93**, 073306 (2008)
16. H. Okamoto, K. Imura, Prog. Surf. Sci. **84**, 199–229 (2009)
17. J. Kerr, Philos. Mag. **3**, 332 (1877)
18. J.B. Gonzalez-Diaz, A. Garcia-Martin, G. Armelles, J.M. Garcia-Martin, C. Clavero, A. Cebollada, R.A. Lukaszew, J.R. Skuza, D.P. Kumah, R. Clarke, Phys. Rev. B **76**, 153402 (2007)
19. G. Armelles, A. Cebollada, A. Garcia-Martin, M.U. Gonzalez, Adv. Opt. Mater. **1**, 10–35 (2013)
20. D. Courjon, *Near-Field Microscopy and Near-Field Optics* (Imperial College Press, London, 2003)
21. J.A. Stroscio, W.J. Kaiser, *Scanning Tunneling Microscopy* (Academic Press Limited, London, 1993)
22. P. Eaton, P. West, *Atomic Force Microscopy* (Oxford University Press, Oxford, 2010)
23. S.I. Bozhevolnyi, V.S. Volkov, T. Sondergaard, A. Boltasseva, P.I. Borel, M. Kristensen, Phys. Rev. B **66**, 235204 (2002)
24. K. Matsuda, T. Saiki, H. Saito, K. Nishi, Appl. Phys. Lett. **76**, 73–75 (2000)
25. H. Heinzelmann, D.W. Pohl, Appl. Phys. A Mater. Sci. Process. **59**, 89–101 (1994)
26. K. Lieberman, A. Harush, A. Lewis, R. Kopelman, Science **247**, 59–61 (1990)

Chapter 7
General Conclusions

Five essential global conclusions can be extracted from this work:

- Magnetoplasmonic interferometry is a very interesting tool to analyze and to measure the magnetic modulation of the SPPs wavevector.
- The parameters that govern the modulation depth of the magnetoplasmonic interferometers are mainly related to the properties of the SPPs, rather than to the magnetooptical properties of the ferromagnetic material. It is specially relevant the role of the electromagnetic field distribution along the materials of the interface, as well as the dispersion relation of the SPPs and how far or close it is from the line of the light ($k_x - k_0$).
- A magnetoplasmonic interferometer can be a device per se. It is a reasonable choice to modulate SPPs. Applying external magnetic fields of 20 mT we have been able to obtain modulation depths of around 2 % on non-optimized Au/Co/Au trilayers, with interferometer sizes on the micro-scale. Moreover, we could obtain modulation depths of about 12 % by substituting Co by Fe, working at around 950 nm and covering the interferometer with a dielectric overlayer of $n_d = 1.49$.
- Plasmonic interferometry is interesting also for biosensing, where it can be more sensitive than traditional SPR sensors. Magnetoplasmonic interferometry offers a new sensing parameter, the magnetic modulation of the SPP, which is more sensitive to refractive index variations.
- The magnetic modulation of SPP can also be obtained from near field experiments. With the appropriate configurations, we could obtain not only the modulation of the SPP wavevector Δk_x, but also that of its vertical component Δk_z.

Below, a more detailed analysis of the conclusions of each part of this thesis can be found:

The first part deals with the spectral dependence of the magnetic modulation of both real and imaginary parts of k_x in noble/ferromagnetic/noble metal trilayers, and the optimization of the system. The spectral evolution of the modulation for both the real and the imaginary parts of the SPP wavevector has been experimentally and theoretically obtained. It has been shown that, although the contribution of the real

D. Martín Becerra, *Active Plasmonic Devices*, Springer Theses,
DOI 10.1007/978-3-319-48411-2_7

part of the SPP wavevector is dominant in most of the analyzed spectral range (500–1000 nm), at longer wavelengths their values approach each other and the imaginary part has to be taken into account to make an appropriate analysis. Both modulations decrease whereas the wavelength increases, and a peak at lower wavelengths is observed in the case of $\Delta k_x{}^r$. This dependence can be qualitatively described by the evolution of the separation between the SPP wavevector and the light line, $k_x - k_0$, which is a relatively simple parameter. This parameter is, nevertheless, a representation of the evolution of the SPP properties with the wavelength, which are related to the vertical confinement of the SPP field. Indeed, the spectral evolution of the electromagnetic field inside the ferromagnetic metal layer has been obtained, being its dependence quite similar to that of Δk_x. Regarding the influence of the ferromagnetic metal, all show the same general behavior, being Fe the one that provides the largest modulation. Nevertheless, if we consider the use of the magnetoplasmonic interferometers as modulators, we should consider not only the magnetic modulation of k_x but also the propagation distance of the SPP, L_{sp}, since the modulated intensity depends on both quantities. Therefore a figure of merit combining both magnetic modulation and propagation distance of the SPP is also analyzed ($2|\Delta_{k_x}| \cdot L_{sp}$). In terms of spectral dependence, the decrease of SPP wavevector modulation is compensated by the increase in L_{sp} for a significant wavelength range, resulting that the 700 nm–1 μm interval becomes the optimal one. Trying to optimize the interferometric system, and using the analytical expression for Δk_x, we have demonstrated that the deposition of a dielectric overlayer on top of our Au/Co/Au multilayers leads to a significant enhancement of the magnetic field induced modulation of the SPP wavevector. In fact, the analysis of the figure of merit shows that the modulation depth of a magnetoplasmonic switch can be increased with the addition of dielectric overlayers despite of the strong reduction in SPP propagation length, which allows to reduce the size of the device.

The second part of this thesis is devoted to evaluate the plasmonic and magnetoplasmonic interferometers as biological sensors. They have been compared to the widely used plasmon-based sensors, known as SPR. Each sensor has been compared in its optimal configuration, and we have considered their sensitivity with the refractive index n of the sensing layer. Moreover, since each method measures different outputs, the sensitivity of the basic physical parameter underlying in each technique has been also determined. It has been observed that the sensitivity of the plasmonic interferometer can be larger than that of the SPR, mainly due to the role of the slit-groove distance in the measured intensity, since they both depend on the same physical parameter ($k_x(n)$). On the other hand, it has also been determined that in the magnetoplasmonic interferometer, the underlying physical parameter $\left(\frac{\delta \Delta k_x}{\delta n}\right)$ is more sensitive than that of the plasmonic interferometer $\left(\frac{\delta k_x}{\delta n}\right)$. However, the measurement procedure used for the magnetoplasmonic interferometer depends on both parameters, $k_x(n)$ and $\Delta k_x(n)$, and although $\Delta k_x(n)$ is more sensitive, it is quite smaller too, resulting that its contribution to the total sensitivity is negligible. Therefore, a measurement procedure able to isolate the evolution of $\Delta k_x(n)$ has to be applied in order to take advantage of the higher sensitivity of magnetoplasmonic interferometers. Regarding the wavelength behavior of the sensitivities, both the SPR and the plasmonic interferometer are more sensitive with increasing wavelength even though the variations of k_x with the refractive index behave in the opposite manner. This is because of the reduced losses of the SPP at higher wavelengths, which generate narrower reflection dips in SPR sensors and allow longer values of d in the plasmonic interferometers. For the ideal MP interferometry (only dependent on $\Delta k_x(n)$), on the other hand, the use of smaller wavelengths would be advantageous as $\frac{\delta \Delta k_x}{\delta n}$, decreases with the wavelength.

The last part of the manuscript shows the possibilities of magnetic modulation in different near field interferometric configurations. The magnetic field induced modulation of k_x could be detected in the near field by measuring the interference of two counterpropagating plasmons between two launching slits. The near field intensity modulation would be proportional to $4\Delta k_x \times d$, being $2d$ the distance between the two slits. It could also be possible to have direct access to the magnetic modulation of k_x^i and even of k_z by measuring the magnetic modulation of a decaying SPP in an isolated slit.

Appendix A
Interferometers: Fabrication, Optical, and Magnetooptical Characterization

A.1 Fabrication of Magnetoplasmonic Interferometers

A.1.1 Layers Deposition

The magnetoplasmonic interferometers consist of Au/Co/Au trilayers with 6 nm of Co and a total thickness of 200 nm. The top Au layer ranges from 5 to 45 nm from the surface (and consequently the bottom one ranges from 189 to 149 nm). They have been grown in an ultra-high vacuum system (10^{-10}) using three Physical Vapor Deposition techniques available at the (IMM): magnetron sputtering, effusion or k-cell evaporation, and (e-beam) evaporation. The layers are deposited over a glass substrate that rotates during the deposition process to homogenize thickness. An initial 2 nm layer of Cr or Ti is included, acting as an adhesion layer between the gold and the glass. Over it, the thick bottom Au layer is deposited, next we add a 6 nm Co (ferromagnetic) layer, and finally all is covered by the thin top Au layer in order to protect the Co layer from oxidation and to provide the good plasmon properties to the system. The complete process is made without taking the sample out of the system.

In principle, in our vacuum system and for our kind of samples, the three available physical vapor deposition techniques could be used for each material. However, due to the quality of the grown material, the effusion cell is used for Au, e-beam evaporation is used for Ti (when needed), and magnetron sputtering is used for both Co and Cr.

After having the materials deposited, the interferometers are engraved by means of (FIB) lithography, also available the facilities of the IMM.

A.1.2 FIB Lithography

FIB lithography refers to milling a predesigned pattern in a material with high resolution by means of a focused ion beam [1–3]. A FIB system basically consists

D. Martín Becerra, *Active Plasmonic Devices*, Springer Theses,
DOI 10.1007/978-3-319-48411-2

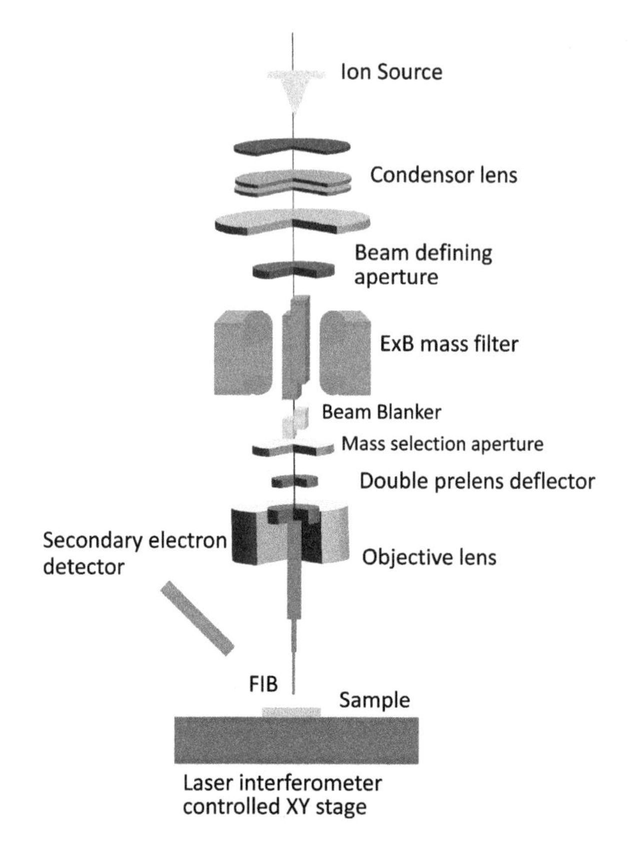

Fig. A.1 Diagram of the central module of a focused ion beam system, with the different elements mentioned in the text

of a vacuum system provided with an ion source [4], a sample stage and a series of electromagnetic lenses. Applying a high voltage between the ion source and an extraction electrode, we produce an ion beam that is focused onto the sample with the electromagnetic lenses. By digitally controlling the voltage applied by the last electromagnetic lenses, this well focused ion current is scanned over the sample following a specified pattern and therefore etching the material with the desired design. A basic diagram of a FIB system is represented in Fig. A.1. The depth of the pattern depends on the ion dose and the sputtering yield of the material. The ion dose is the charge per unit area applied to the sample, and is controlled by the applied ion current and the exposure time. Regarding the sputtering yield of a material, the higher it is, the less dose is needed to etch the same depth. Therefore, each time a new material or combination of materials is used, a calibration process for the dose (mC/cm^2) needed to etch the desired depth has to be done. The main advantage of using FIB lithography for the patterning is that it is a direct lithography method, i.e. it does not need any resist. Thus different milling depths can be achieved in the same process, just by specifying the suitable dose for each component of the total design. However, since the etching is obtained by sputtering, i.e. by hitting the sample with the source ions, this method introduces lots of defects as well as implantation of

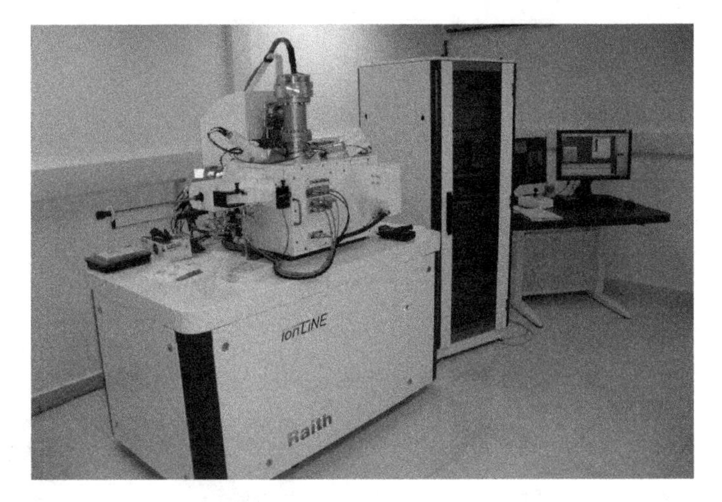

Fig. A.2 Photograph of the IMM focused ion beam system

some of the ions on the sample. This may limit the use of FIB lithography for certain applications. Another restriction arises from charging effects. As electrons are extracted from the sample when the ions hit it, the addition of a conducting layer is needed to work with isolating samples. The electrons emitted when the ion beam is scanned over the sample allow to image it by means of a secondary electron detector. However, an intensive use has to be avoided, since this imaging process takes place while the sample is being etched. To allow simultaneous FIB etching and imaging of the sample, there also exist dual systems with both electron and ion sources and column.

Except for the very first interferometers that were patterned at Konstantz University, all the other interferometers have been patterned at the IMM with an Ion-Line system from Raith (Fig. A.2). This is a FIB system with a single 30 kV ion column and Ga as the ion source. In order to select the most appropriate current to engrave the wanted pattern, a set of different apertures installed within the column is used (beam defining aperture in Fig. A.1). Those apertures range from 5 μm diameter to 1 mm. For the fabrication of our interferometers, we usually use the 20 or 30 μm apertures, which provide a 8–16 pA of current intensity, and a spot size of about 15–30 nm. This spot size allows us to obtain slits and grooves of around 100 nm width. Once the ion current is fixed by the selected aperture, we will achieve the desired dose by setting the exposure time. For each material or combination of materials used, we have first calibrated the dose needed to etch the desired depth. Figure A.3 shows the calibration process. A series of micron-sized squares with different doses are patterned on the material using FIB lithography. Then the depths of those squares are measured by atomic force microscope (AFM), and the results are plotted to extract the dose needed to mill the desired depth of that material. In particular, a required total dose of 12 and 5.5 $\mathrm{mC/cm^2}$ is applied to engrave the slit and the groove respectively in our trilayers. Although the calibration process is usually made at the microscale,

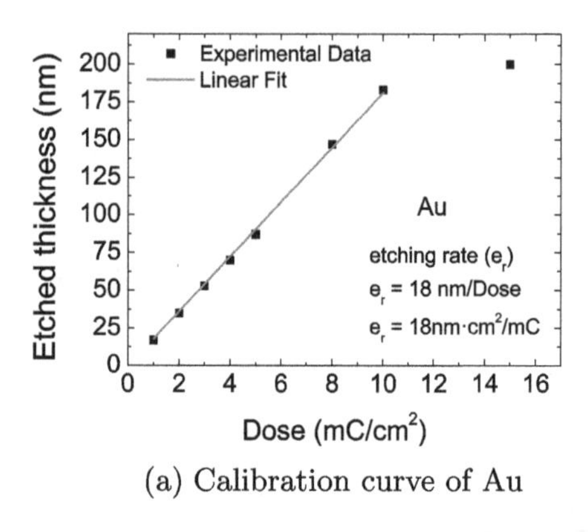

(a) Calibration curve of Au

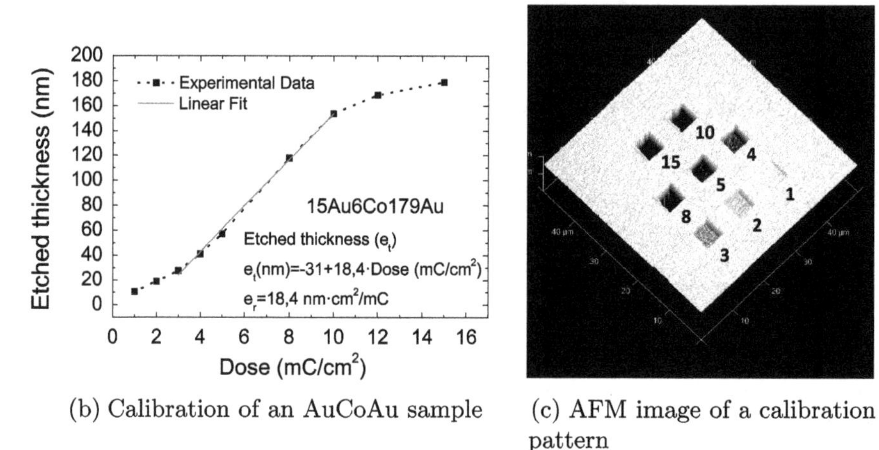

(b) Calibration of an AuCoAu sample

(c) AFM image of a calibration pattern

Fig. A.3 Calibration process. **a** Calibration curve for a pure Au sample with the linear fit where the slope shows the etching rate, i.e. nm of depth milled per mC/cm^2, which in this case is 18 nm per cm^2/mC. The last experimental spot is not fitted since there we are already etching the glass substrate, which is a different material and therefore it presents a different etching rate, **b** calibration curve for a 15Au/6Co/179Au sample. The different slopes correspond to the different materials. The linear fit of the central part of the curve, corresponding to the thick Au layer, shows the same etching rate as a pure Au layer, as expected. The full linear equation gives in this case the total etched thickness per mC/cm^2, **c** AFM image of the calibration pattern for the Au layer, where it can be seen several micron-sized squares with different applied dose ($\times$1000 mC/cm^2)

those dose values can be applied to fabricate different structures at the nanoscale, as Fig. A.4 shows. In Fig. A.4, two groups of parallel slits with different doses but the same width (120 nm) in a 200 nm Au layer are shown. In order to be able to see the slits profile those images are done with the sample tilted 50°, besides, a completely milled rectangle is shown as a reference (in the images, the complete black area). The upper part of the image shows the Au layer, and the lower part corresponds to a completely exposed glass area. As it can be seen, for a dose of 3.5 mC/cm^2, the slits patterned in the Au layer do not cross completely the material, while for the

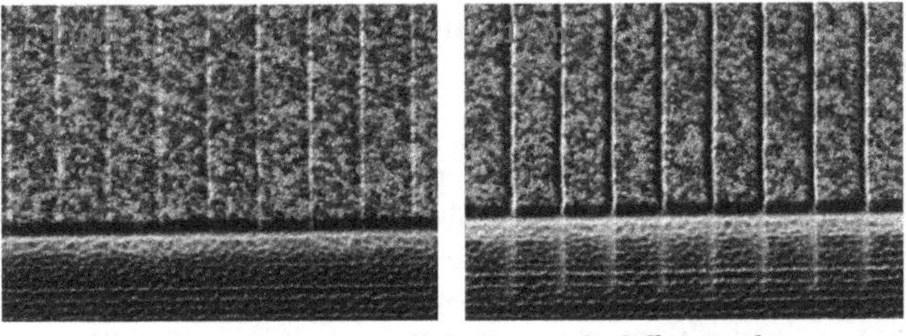

(a) FIB image of parallel slits with different doses

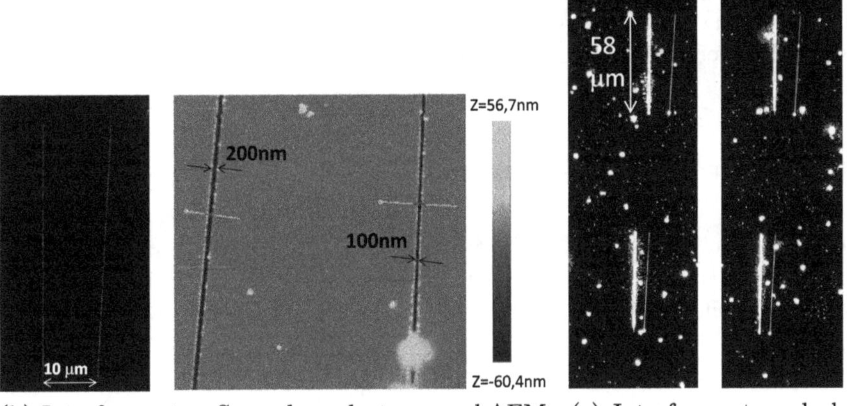

(b) Interferometers Secondary electrons and AFM images (c) Interferometers dark field images

Fig. A.4 **a** FIB (secondary electrons) image of parallel slits with different doses, the *left* one is made with an applied dose of 3.5 (60 nm etched approximately) while the ones of the *right* are fabricated with an applied dose of 12, enough to etch all the material, **b** at the *left* there is a secondary electrons image of an interferometer of 48 µm length and 3° and 10 µm of initial separation between slit and groove, while on the *right* it is shown an AFM image of a similar interferometer, **c** dark field (DF) image of different interferometers. The *left image* shows interferometers of 58 µ length and 1° (*up*) and 2° (*bottom*) with 10 µm of initial separation between slit and groove. The *right images* show a DF image of an interferometer of 58 μ length and 2° and 10 (*up*) and 20 µm (*bottom*) of initial separation. All these interferometers are engraved in a 15Au/6Co/179Au sample

12 $\mathrm{mC/cm^2}$ dose, the slits, of nanometric dimensions, also cut through the metal area.

Different interferometer designs have been fabricated, such as tilted slit-groove couples and parallel slits, with different depths and widths. An example of those is shown in Fig. A.4. There, an AFM image and the sizes of one of the analyzed pairs of interferometers is shown. A dark field image of other analyzed interferometers with different angles and initial distances is shown also in Fig. A.4. All the defects that appear in this last are typical of ours metallic multilayers.

A.2 Optical, Magnetooptical, and Magnetic Characterization

In order to analyze completely and in a proper way the magnetic modulation of the interferometers, we need to know the optical and magnetooptical (MO) constants, as well as some magnetic parameters of the materials that we use. Although general values of the optical constants for bulk material can be seen in the literature or in well established handbooks [5–7], the values for thin layers depend on the growth conditions. So to get a better description of our system, we have determined them experimentally. The optical constants have been determined by ellipsometry and the MO ones by Polar Kerr spectroscopy. Finally, we characterize the samples magnetically by using the Transverse Kerr effect, taking into account that the MO activity is (as a first approximation) linearly dependent with the magnetization.

A.2.1 *Spectroscopic Ellipsometry*

As it has been said in Sect. 2.3, the macroscopic description of the electromagnetic response of a material to incident light is given by the dielectric tensor. Usually this tensor, without considering any magnetooptical effect, is diagonal and can be written as:

$$\begin{pmatrix} \varepsilon_{xx} & 0 & 0 \\ 0 & \varepsilon_{yy} & 0 \\ 0 & 0 & \varepsilon_{zz} \end{pmatrix}, \tag{A.1}$$

where ε_{xx}, ε_{yy}, and ε_{zz} are the dielectric constants of the material along the different axis. For an isotropic material, this can be simplified to:

$$\begin{pmatrix} \varepsilon & 0 & 0 \\ 0 & \varepsilon & 0 \\ 0 & 0 & \varepsilon \end{pmatrix}, \tag{A.2}$$

where ε is the dielectric constant, and is generally complex. Ellipsometry is a well known optical technique to obtain the dielectric constants of a material as a function of the wavelength [8]. It does not require any special sample preparation, and it is a non destructive technique. It compares the changes on the polarization state of the light reflected from the sample with those of the incident light. The obtained results are fitted with a model, usually assuming a homogeneous layer of the material on a substrate or a multilayer stack. The fitting uses several parameters of the sample (refractive index of the different materials, thicknesses, roughness, anisotropy...). Therefore, the best results are obtained if we can provide the maximum information about our sample obtained by independent methods, such as layers thicknesses, surface roughness, interface roughness, etc. This method can be used to determine the

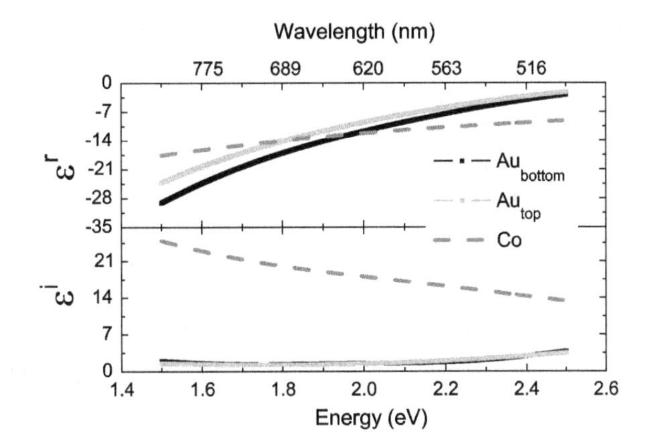

Fig. A.5 Dielectric constants for the three layers of the $25Au/6Co/169Au$ trilayers used in the studied MP interferometers. *Upper graph* real part of the dielectric constants. *Lower graph* imaginary part of the dielectric constants

optical dielectric constants of the material, but it can also provide layer thicknesses, roughness, and other parameters. Deeper analysis of how ellipsometry works, and the different models that can be applied, can be found at [8].

The ellipsometer used at the IMM is a M200FI J. A, Woollam Co.^{TM}, where the angle of incidence can be changed from 45° to 75°, and the wavelength ranges from about 500 nm to 1.69 μm. With it, we have determined the optical constants for the different layers (Au_{top}, Au_{bottom} and Co) for our magnetoplasmonic trilayers. The obtained values are shown at Fig. A.5. Due to their different thicknesses, and the different base materials and the roughness that they introduce, the constants of the top Au layer and the bottom Au layer are not the same, although they are quite similar (Fig. A.5). We observed that all trilayers provided similar dielectric constants for our materials, so we have used the ones obtained from the trilayer with 25 nm Au on top for all the simulations. As an example, in Sect. 4.2, we study the enhancement of the magnetic modulation by using a dielectric with a higher refractive index. The experimental demonstration is performed by coating PMMA on top of our interferometers. The thickness of PMMA was assessed by ellipsometry, fitting just the PMMA thickness, providing all the other thicknesses and optical constants (top Au, Co, and bottom Au) as fixed parameters to the fitting, since they had been previously determined. The resulting PMMA thickness was of 60 nm, which agreed with the previously made calibration tests.

A.2.2 Transverse Kerr Magnetic Characterization

An important parameter for our MP interferometers is the magnetic field at which the Co layer is saturated.In order to determine it, we measured the magnetic hysteresis loops when the magnetic field is applied within the sample surface plane (the relevant configuration for our magnetoplasmonic study) by means of the transverse magnetooptical Kerr effect (TMOKE). As it has been mentioned in Sect. 2.3, when a magnetic field is applied parallel to the sample plane (XY plane) but perpendicular to the plane of incidence (XZ plane, Fig. 2.8), a non diagonal dielectric component will appear, in this case ε_{xz}. This produces a change in the intensity of the p-polarized reflected light, without any change of the polarization, known as transverse MO Kerr effect. Therefore, in transverse Kerr configuration, the reflection coefficient of p-polarized light, $r_{pp} = r$, depends on the sample magnetization:

$$r(B) = r_0 + a \cdot m, \tag{A.3}$$

where r_0 is the reflectivity coefficient when there is no magnetic field ($B = 0$), a is a parameter which depends on the sample structure and the material and m is the magnetization of the sample normalized to the magnetization at saturation ($m \in [-1, 1]$). The transverse Kerr effect is usually represented by the relative variation of reflectivity:

$$\begin{aligned} R(B) &= |r(B)|^2, \\ TMOKE \equiv \frac{\Delta R}{R} &= \frac{R(B^+) - R(B^-)}{R(B^+) + R(B^-)} \approx \\ &\approx \frac{R(B^+) - R_0}{R_0} = \frac{2r_0 am}{r_0^2}, \end{aligned} \tag{A.4}$$

being the superindex $^+$ or $^-$ the sign of the magnetic field (and therefore the magnetization) applied. TMOKE effect depends on the optical and magnetooptical constants of the involved materials through r_0 and a. As it also depends, and linearly, of m, it allows to obtain magnetic hysteresis loops of the material. The hysteresis loops of our trilayers, for a magnetic field applied along the sample surface, have been obtained with a TMOKE setup installed at the IMM. The magnetic field is applied in a given direction, increasing it by small steps up to a maximum, and then it is decreased, then we change the sign of the applied magnetic field and make the same, completing the loop. With a laser and a series of detectors and polarizers, the reflectivity as a function of the magnetic field applied is measured. The TMOKE signal $\left(\frac{\Delta R}{R}\right)$ is calculated at each value of the magnetic field applied [9], and the obtained loop represents the evolution of the magnetization with the magnetic field (see Eq. A.4).

The obtained loop characteristic of all the trilayers (Fig. A.6), is a square one with rounded edges, which is typical from polycrystalline Co layers of this thickness [10, 11]. The saturation field, which is the magnetic field at which the sample is magnetically saturated is quite low, of the order of 15–20 mT. This is the most relevant parameter for the measurements of the interferometers, since they have to be saturated.

When determining Δk_x with the MP interferometers, it is essential to work with the sample saturated. In this way, the magnetic effect will be the largest one, and the results will be stable and reproducible. To double cross-check the samples were saturated during the experiments, besides the determination of the magnetic hysteresis loops, we also performed the following measurement: we measured the amplitude of

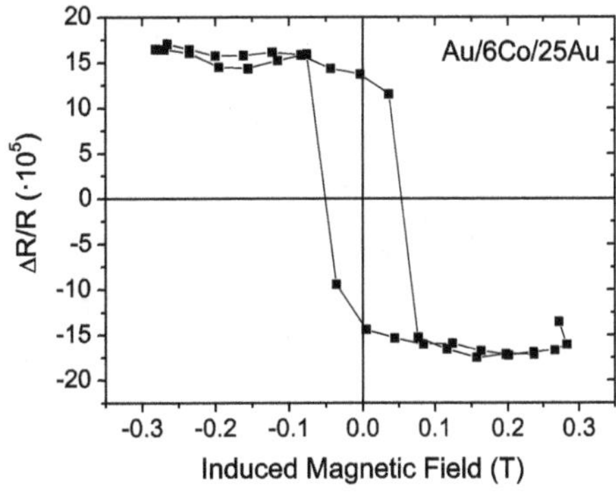

Fig. A.6 TMOKE signal for a 25Au/6Co/169Au magnetoplasmonic trilayer

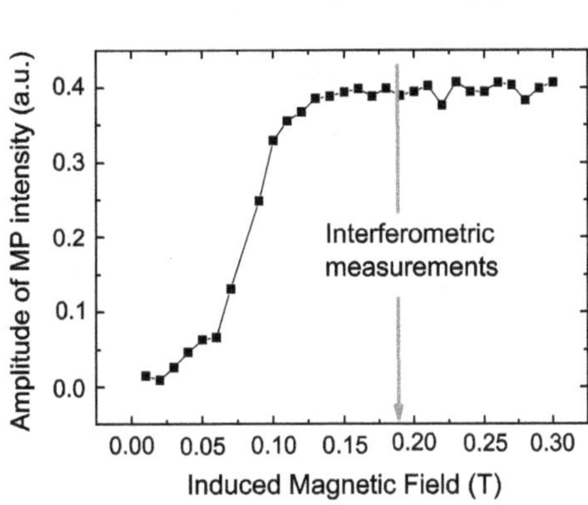

Fig. A.7 Amplitude of the MP intensity of the interferometer as a function of the applied magnetic field

the MP intensity for a given portion of the slit of our interferometer as we increased the applied magnetic field. As it is shown in Fig. A.7, the value of saturation agrees with the one obtained from the hysteresis loop (Fig. A.6).

Finally, to mention that the TMOKE effect can also be directly used to obtain the magnetic modulation of the wavevector of a SPP using a Krestchmann configuration [10]. Nevertheless, we use it to obtain hysteresis loops of our trilayers.

A.2.3 Magnetooptical Characterization: Polar Kerr Spectroscopy

We obtained the magnetooptical constants of our ferromagnetic material by means of a polar Kerr spectrograph already installed in our laboratory. As it is mentioned in Sect. 2.3, in the case of polar configuration the magnetization is perpendicular to the sample plane and in the plane of incidence (Figs. A.8 and 2.8). In this situation, it is known that the magnetooptical constants that appear in the dielectric tensor are ε_{xy} (Eq. A.5). This means that, when p-polarized light is reflected through the sample under that magnetic field, the reflected light will show a small component of s-

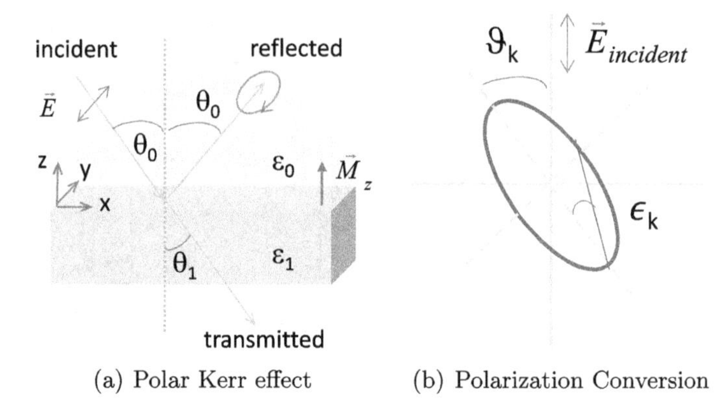

Fig. A.8 **a** Polar Faraday (transmission) and Kerr (reflection) effects. ε_0 represents the "non-magnetic" material and ε_1 the "magnetic" one, **b** description of the polarization conversion that takes place in this configuration

polarized light and viceversa. The reflected light becomes then elliptically polarized, with its major axis rotated from its initial incident polarization plane.

$$\begin{pmatrix} \varepsilon_{xx} & \varepsilon_{xy} & 0 \\ -\varepsilon_{xy} & \varepsilon_{yy} & 0 \\ 0 & 0 & \varepsilon_{zz} \end{pmatrix} \tag{A.5}$$

The characteristic parameters that define this polar magnetooptical Kerr effect (PMOKE) are then the rotation ϑ_k and the ellipticity ϵ_k that characterize the elliptically polarized reflected light. These ellipticity and rotation are proportional to the magnetization M_z [12]. For the most simple situation, which is a semi-infinite non magnetic material, and a semi-infinite ferromagnetic one, those parameters can be written as:

$$\begin{aligned} \vartheta_k + i\epsilon_k &= -\frac{r_{sp}}{r_{pp}} = \frac{n_0 \cos \vartheta_0 \varepsilon_{xy}}{n_1(n_1^2 - n_0^2)\cos(\vartheta_0 + \vartheta_1)} \\ &\approx \frac{\epsilon_{xy}}{n_1(1 - n_1^2)}, \end{aligned} \tag{A.6}$$

where ϑ_0 and ϑ_1 are the angles of light within the medium 0 and the medium 1 respectively. n_0 and n_1 are the refractive indexes of the "non-magnetic" and "ferromagnetic" materials. The approximation is valid when the incident medium is vacuum and at normal incidence. As it can be seen, from this equation, by measuring the rotation and the ellipticity and knowing the optical constants of the material, the magneto-optical constants can be determined. The PMOKE Spectrograph at the IMM allows to determine the rotation and the ellipticity as a function of the wavelength and the applied magnetic field by using a set of polarizers and analyzers and a photoelastic modulator [9, 13]. During the measurements, it is very important that we work in

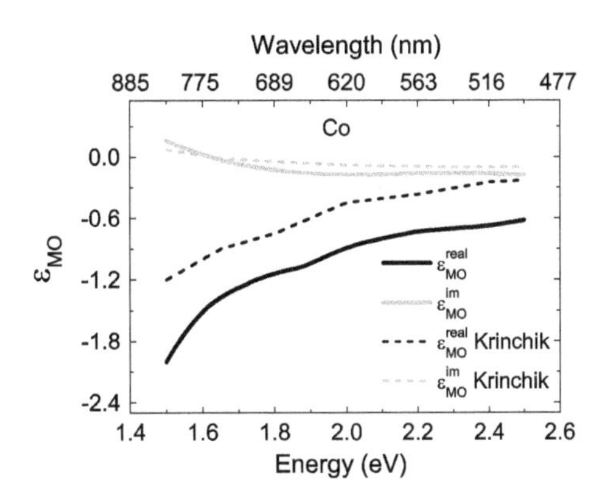

Fig. A.9 Experimental and Krinchik [14] magnetooptical constants of Co. The imaginary part values are very similar, while our experimental real part is larger than that from the literature

saturation conditions, that are different than those of the transverse configuration (we have made this measurements applying 1.6 T, although saturation is achieved at about 1 T [11]). Having previously obtained the optical constants of our layers, we can then calculate the magnetooptical constants of the magnetic material (Co in our case). In Fig. A.9, the experimentally determined values of the MO constants for our Co layers are shown and compared to some values found in the literature [14].

References

1. H. Morimoto, Y. Sasaki, K. Saitoh, Y. Watakabe, T. Kato, Microelectron. Eng. **4**, 163–179 (1986)
2. C.A. Volkert, A.M. Minor, MRS Bull. **32**, 389 (2007)
3. M.. Villarroya-Gaudo, *Diseño y fabricacion de sistemas micro/nano electromecanicos integrados monoliticamente para aplicaciones de sensores de masa y sensores biologicos con palancas como elementos transductores*. Ph.D. thesis (Universitat Autonoma de Barcelona, 2005)
4. R. Muhle, Rev. Sci. Instrum. **63**, 3040–3049 (1992)
5. E.D. Palik, *Handbook of Opical Constants of Solids* (Academic Press, San Diego, CA, 1998)
6. P.B. Johnson, R.W. Christy, Phys. Rev. B **9**, 5056–5070 (1974)
7. P.B. Johnson, R.W. Christy, Phys. Rev. B **6**, 4370–4379 (1972)
8. H.G. Tompkins, E.A. Irene, *Handbook of Ellipsometry* (William Andrew Publishing, Norwich, NY, 2005)
9. E. Ferreiro-Vila, *Intertwined Magnetooptical and Plasmonic Properties in Metal and Metal/Dielectric Magnetoplasmonic Multilayers*. Ph.D thesis (Universidad de Santiago de Compostela, 2012)
10. E. Ferreiro-Vila, M. Iglesias, E. Paz, F.J. Palomares, F. Cebollada, J.M. Gonzalez, G. Armelles, J.M. Garcia-Martin, A. Cebollada, Phys. Rev. B **83**, 205120 (2011)
11. E. Ferreiro-Vila, J.B. Gonzalez-Diaz, R. Fermento, M.U. Gonzalez, A. Garcia-Martin, J.M. Garcia-Martin, A. Cebollada, G. Armelles, D. Meneses-Rodriguez, E. Muoz, Sandoval. Phys. Rev. B **80**, 125132 (2009)
12. J.L. Erskine, E.A. Stern, Phys. Rev. B **8**, 1239–1255 (1973)
13. W.S. Kim, M. Aderholz, W. Kleemann, Meas. Sci. Technol. **4**, 1275 (1993)
14. G.S. Krinchik, J. Appl. Phys. **35**, 1089–1092 (1964)

CPSIA information can be obtained
at www.ICGtesting.com
Printed in the USA
LVHW02*1748180318
570251LV00001B/9/P

9 783319 484105